T0075326

Learning and Analytics in Intelligent Systems

Volume 6

The main aim of the series is to make available a publication of books in hard copy form and soft copy form on all aspects of learning, analytics and advanced intelligent systems and related technologies. The mentioned disciplines are strongly related and complement one another significantly. Thus, the series encourages cross-fertilization highlighting research and knowledge of common interest. The series allows a unified/integrated approach to themes and topics in these scientific disciplines which will result in significant cross-fertilization and research dissemination. To maximize dissemination of research results and knowledge in these disciplines, the series publishes edited books, monographs, handbooks, textbooks and conference proceedings.

More information about this series at http://www.springer.com/series/16172

D. Jude Hemanth
Editor

Human Behaviour Analysis Using Intelligent Systems

Editor
D. Jude Hemanth
Department of Electronics and
Communication Engineering
Karunya University
Coimbatore, Tamil Nadu, India

ISSN 2662-3447 ISSN 2662-3455 (electronic)
Learning and Analytics in Intelligent Systems
ISBN 978-3-030-35138-0 ISBN 978-3-030-35139-7 (eBook)
https://doi.org/10.1007/978-3-030-35139-7

This Springer imprint is published by the registered company Springer Nature Switzerland AG
The registered company address is: Gewerbestrasse 11, 6330 Cham, Switzerland

Contents

Real Time Palm and Finger Detection for Gesture Recognition Using Convolution Neural Network

R. Ramya and K. Srinivasan

Abstract Hand motion is one of the methods for communicating with the PC to perform colossal application. The goal is to apply the machine learning algorithm for quicker motion route acknowledgment. This application pursues the learning of motions and recognizing them accurately. At first Convolution Neural Network (CNN) model is prepared with various pictures of various hand signals for different people for grouping the letters in order. The initial step is realizing, where the data is stacked and in pre-handling step, edge, mass locators, twofold thresholding, limit box recognitions are used to extract the features. By utilizing ConvNet, which is a machine learning algorithm, the input picture's features are cross confirmed. The accuracy is found to be 98% and this methodology is effective to address the impact of various problems. The assessment of this strategy shows to performing hand gesture acknowledgment.

Keywords Image processing · Convolution Neural Network · Deep Learning · Gesture identification

1 Introduction

In the advanced world, communication between the individuals assumes a fundamental job for the improvement of the nation. In spite of the fact that there are numerous strategies for the tragically challenged correspondence despite everything it slacks a ton in investigating the considerations. In this task, the proposed strategy propels the correspondence of the tragically challenged by presenting the Hand Gesture Recognition (HGR). In spite of the fact that numerous calculations are created for acknowledgment, this paper gives the best technique with more

R. Ramya (✉) · K. Srinivasan
Department of EIE, Sri Ramakrishna Engineering College, Coimbatore, India
e-mail: ramya.r@srec.ac.in

K. Srinivasan
e-mail: hod-eie@srec.ac.in

D. J. Hemanth (ed.), *Human Behaviour Analysis Using Intelligent Systems*,
Learning and Analytics in Intelligent Systems 6,
https://doi.org/10.1007/978-3-030-35139-7_1

precision for acknowledgment. In this, HGR is finished by python where numerous libraries like OpenCV, TensorFlow, Keras and so on, are utilized to process and order the picture with the assistance of the Deep Learning (DL) idea. There are numerous calculations which are utilized for recognizing the fingers. Shaikh et al. [1] advises the approaches to offer sign to PC vision gadgets by utilizing hand motion. All the more explicitly hand motion is utilized as the sign or info methodology to the PC. These will profit the whole client without utilizing an immediate gadget and can do what they need as long as the PC vision gadget can detect it. These make PC client simpler than utilizing the console or mouse. The condition of specialty of human PC connection shows the realities that for controlling the PC procedures motions of different kinds of hand developments have been utilized. The particulars additionally include precision of location and acknowledgment for our application introduces an increasingly compelling and easy to use techniques for human PC communication brilliantly with the use of hand gestures. Panwar and Mehra [2] proposes about the advancement of another sort of Human Computer Interaction framework that curbs the issues that clients have been looking with the present framework. The task is executed on a Linux framework yet could be actualized on a windows framework by downloading a few modules for python. The calculation applied is impervious to change in foundation picture as it did not depend on foundation picture subtraction and isn't modified for a particular hand type; the calculation utilized can process diverse hand types, perceives number of fingers, and can complete undertakings according to prerequisite. As it is expressed inside this paper, the principle objectives were come to. The application is fit for the signal acknowledgment progressively. There are a few confinements, which despite everything we must be defeated in future. Ray et al. [3] procedure for human PC cooperation utilizing open source like python and OpenCV the hand signals are taken by a camera. Picture changes are done on the rgb picture to change over into ycbcr picture. The ycbcr picture changed into twofold picture. This calculation needs uniform and plane foundation. Edge identification calculation is utilized to discover the edges in the picture. By utilizing edge discovery the direction of hand is recognized. The highlights like centroid, tops recognition, Euclidean separation and thumb location are found. In this paper, they have considered five bits to speak to the hand picture. That is first piece speaks to whether the thumb is available or not. On the off chance that it is available, the bit is given as 1 else 0. The remaining four bits represents to the four fingers. The achievement rate is 92% with calculation time 2.76 s. The algorithm is executed in MATLAB.

Now a days, numerous collaboration techniques are produced for the correspondence among Human and Human and among Human and Machines and this is known as the Human Machine Interface (HMI). Signal Recognition is likewise one of the techniques for cooperation. This strategy most likely compelling for hard of hearing, idiotic and notwithstanding for visually impaired (for some degree) people for correspondence between two man or notwithstanding for HMI. These individuals will be felt simple to speak with others by signal, at that point composing or composing their musings. In Existing approach the signals acknowledgment is done before just with the handling of current picture and the idea of DL isn't utilized. Thus, it sets aside a long effort to foresee the signal and the precision additionally low. Because of that, it isn't perceived as a decent technique for correspondence between individuals. It is primarily bound on the explanation that; DL isn't renowned and the strategies and methods are believed to be confused and DL is just in the exploration region at those days. Along these lines, DL isn't utilized and the calculations are seen as confused and the exactness additionally not in the best level. The devices utilized for the DL is likewise not referred to, for example, Python, MATLAB, R language and so forth, are not utilized for expectation of signal. In this way, the most punctual calculations are seen as muddled and are not utilized for ongoing application.

2 Related Background

In this project, the images of hands are acquired with the help of the Webcam of the PC. Thus, the live video is processed for HGR. Here, the hand is placed in the box which is viewed in the PC. Then, the hands are detected and then it is captured as a snapshot for further processes. Here, hundreds of datasets are prepared for training. They are trained and tested for accuracy. The more the datasets, the more the accuracy. In these datasets, all the features are extracted using CNN (Convolution Neural Network). Here, many hidden layers are created for feature extraction. The hidden layers are the layers which can't be seen. These are also called as the processing layer. Here, the DL is performed with the help of CNN. The Neural Network is a concept which is similar to that of the brain which consists of neurons. Here, the input and output are known as layers. There are many datasets which is chosen as input and only one output is derived as a result of the Neural Networks.

The representation of Neural Network is shown in Fig. 1. Here, the many inputs are passed as a layer to the hidden layers. They are processed and then transmitted to the next hidden layers up to the output layer. Here, the output will be one, such that, all the features belongs to the same one. That is the variation in input is trained for the detection. The next important part is the weight detection. That is, the output is the only part of the whole system. So, the user has to initialize the weight according to the need. The bias is also added to the output along with the weight to make the system constant. Thus, the bias is a constant independent of input.

The weight along with the bias is represented as Neural Network as shown in Fig. 2. Here, the hidden layer has two inputs. Each input has different weights and same bias. Hence, the equation can be written as,

$$y = W1 * x1 + W2 * x2 + b \quad (1)$$

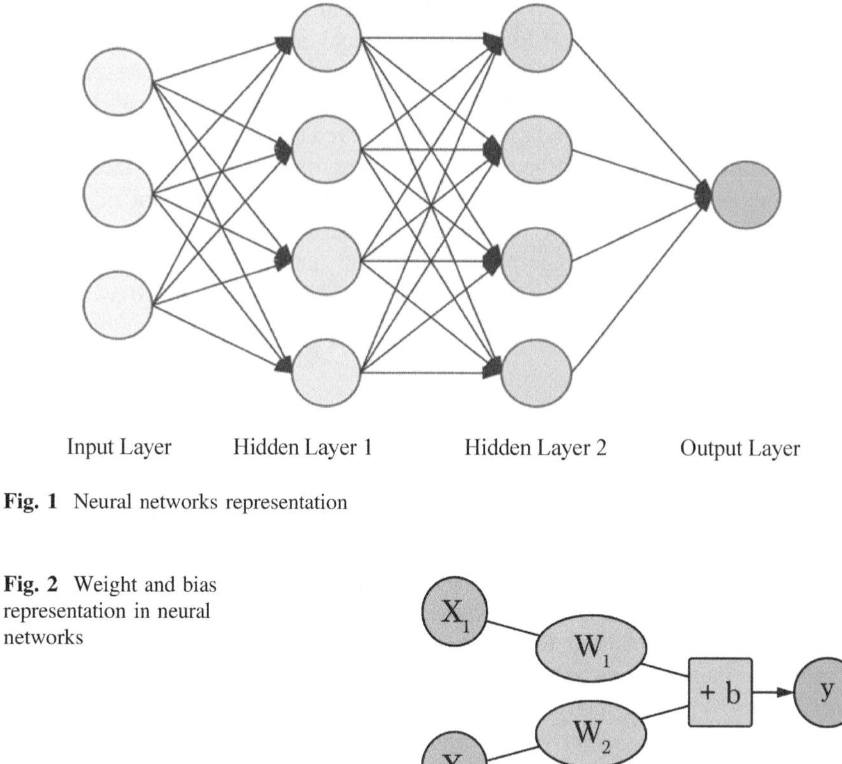

Fig. 1 Neural networks representation

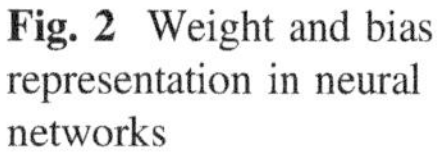
Fig. 2 Weight and bias representation in neural networks

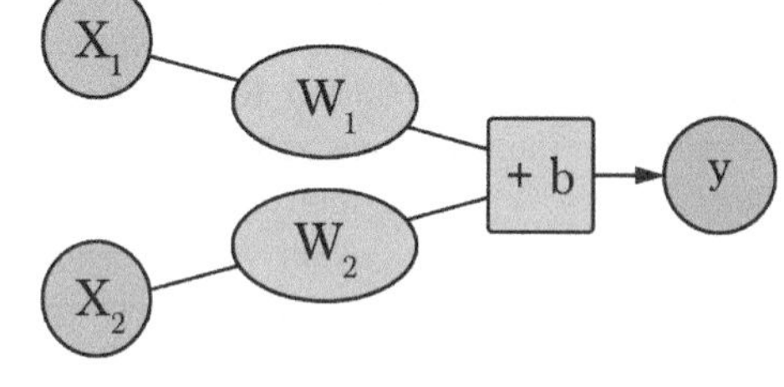

In the above Eq. (1), the ‘y’ is the output, ‘x1’ and ‘x2’ are the inputs, ‘W1’ and ‘W2’ are the weights and ‘b’ is bias. The input may be an image, where only the specific region is selected for feature extraction which is done with the help of the weights.

This is the most basic operation for DL. These processes can be done with the help of many tools such as MATLAB codes, Python codes etc., this paper explains the Python tool for HGR. Chitade and Katiyar [4] uses the K-means clustering algorithm to partition the input image for segmentation. They make use of bounding box to find the orientation. Features like centroid, Euclidean distance are measured for detection. Here the hand is represented by making use of seven bits. First bit represents the orientation of the hand. Second bit is for the presence of thumb in the finger and next three bits are for representing the number of fingers raised. Last two bits for differentiating the gestures which have equal number of fingers. This algorithm has a success rate of 94% with computation time 0.60 s. The algorithm is implemented in MATLAB. Guan and Zheng [5] gives an algorithm for non-uniform background or 3D complex space. Here author will make use of HMM based method to recognize the hand gestures with non-uniform background. The input images are taken by a camera. Skin colour is used for segmentation.

The gestures are splitted by making use of spotting algorithm. They use data aligning algorithm to align features with success rate of 100%. This will give an idea about image processing in an non-uniform background. Stergiopoulou and Papamarkos [6] Proposed vision-based, glove-based and depth-based approach for hand gesture recognition. However, hand gesture itself is simple and a natural way to interact. In otherwise, hand gesture recognition using finger tracking and identification can be implemented more robust and subtle recognition. Recently, new horizons are open with the development of sensors and technology such as Kinect and Depth-Sense. This development has made it possible for robust recognition, like finger identification and hand gesture recognition in bad conditions such as dark light and rough background as well. In this paper, we proposed a new finger identification and hand gesture recognition techniques with Kinect depth data. Our proposed finger identification and gesture recognition methods provide natural interactions and interface by using fingers. We implemented interfaces and designed hand gestures using this method. This paper explains finger identification method and hand gesture recognition in detail. We show the preliminary experiment for evaluating accuracy of finger identification and hand gesture recognition accuracy. Finally, we discuss the result of evaluation and our contributions. Herbon et al. [7] introduces a hand gesture recognition sensor using ultra-wideband impulse signals, which are reflected from a hand. The reflected waveforms in time domain are determined by the reflection surface of a target. Thus, every gesture has its own reflected waveform. Thus, we propose to use machine learning, such as convolutional neural network (CNN) for the gesture classification. The CNN extracts its own feature and constructs classification model then classifies the reflected waveforms. Six hand gestures from American sign language (ASL) are used for the experiment and the result shows more than 90% recognition accuracy. For fine movements, a rotating plaster model is measured with 10° step. An average recognition accuracy is also above 90%.

3 Convolution Neural Network

Convolution Neural Network also called as CNN or ConvNet. It is one of the types of Neural Network which is expertise in processing data which are in the form of grid like topology. That is, Frames or Images. All the images can be represented in the binary form and the CNN will process the image in its binary form. That is, every image will be in the grid of different values. These values are the pixel values that is, the value corresponds to the brightness and the colour of the image.

The binary pixel value representation of the coloured image is shown in Fig. 3. For humans, brain will decide the letter in the image. For that, numerous neurons will work and all the features of the image is determined and the particular feature is taken for prediction. Likewise, each neuron in the CNN layer will detect the simpler patterns such lines, curves etc., and then detect the complex pattern such as face, object etc. Thus, CNN will enable sight for the Computers. An architecture of

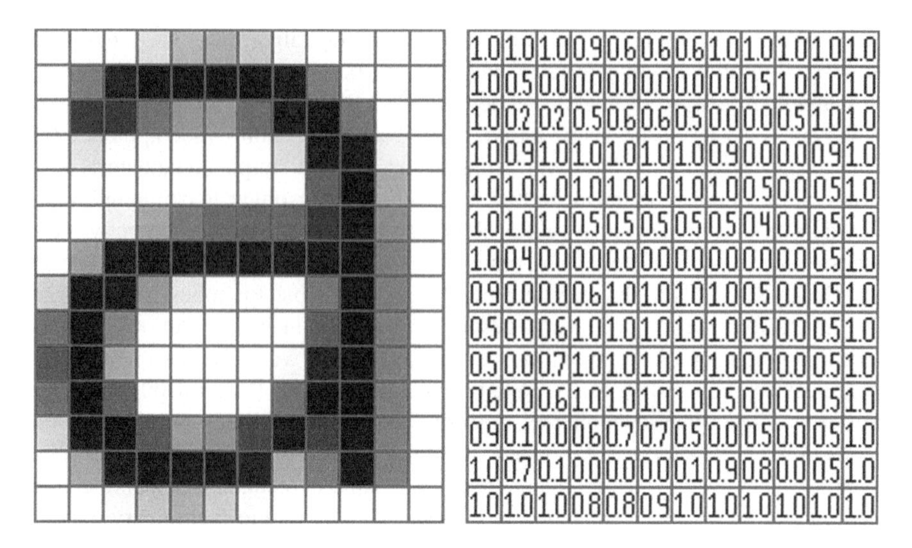

1.0	1.0	1.0	0.9	0.6	0.6	0.6	1.0	1.0	1.0	1.0	1.0
1.0	0.5	0.0	0.0	0.0	0.0	0.0	0.0	0.5	1.0	1.0	1.0
1.0	0.2	0.2	0.5	0.6	0.6	0.5	0.0	0.0	0.5	1.0	1.0
1.0	0.9	1.0	1.0	1.0	1.0	1.0	0.9	0.0	0.0	0.9	1.0
1.0	1.0	1.0	1.0	1.0	1.0	1.0	1.0	0.5	0.0	0.5	1.0
1.0	1.0	1.0	0.5	0.5	0.5	0.5	0.5	0.4	0.0	0.5	1.0
1.0	0.4	0.0	0.0	0.0	0.0	0.0	0.0	0.0	0.0	0.5	1.0
0.9	0.0	0.0	0.6	1.0	1.0	1.0	1.0	0.5	0.0	0.5	1.0
0.5	0.0	0.6	1.0	1.0	1.0	1.0	1.0	0.5	0.0	0.5	1.0
0.5	0.0	0.7	1.0	1.0	1.0	1.0	1.0	0.0	0.0	0.5	1.0
0.6	0.0	0.6	1.0	1.0	1.0	1.0	0.5	0.0	0.0	0.5	1.0
0.9	0.1	0.0	0.6	0.7	0.7	0.5	0.0	0.5	0.0	0.5	1.0
1.0	0.7	0.1	0.0	0.0	0.0	0.1	0.9	0.8	0.0	0.5	1.0
1.0	1.0	1.0	0.8	0.8	0.9	1.0	1.0	1.0	1.0	1.0	1.0

Fig. 3 Binary representation of RGB image (pixel representation)

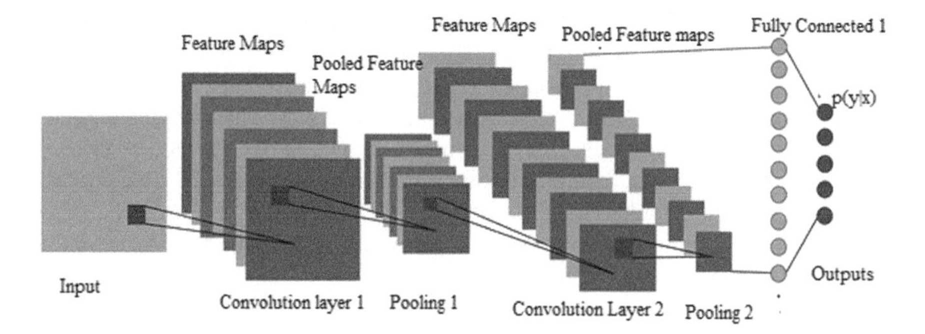

Fig. 4 Architecture of CNN layers

Convolution Neural Network is shown in the (Fig. 4). It is a class of DNN, it helps to analyze the visual imaging.

Layers are the basis for CNN. They generally perform dot product between two matrices. Here, one matrix consists of learning parameters and other consists of restriction portion. In this, the first matrix also called as kernel which is very small when compared to the image. But, it has a depth as that of the image. That is, each image consists of three layers such as Red, Green, Blue. The kernel will contain all the three channels. But it will be less when compared to that of a real image. During its forward pass, the kernel will slide across in both horizontal and vertical direction. That is, along length and width. This produces the 2D representation called activation map. The slide is known as stride and based on this, the spatial position of the image can be calculated.

If the image is of size $W \times W \times D$ and D_{out} number of kernel with special size of F with strides S and amount of padding P, then the size of output volume is given by,

$$W_{out} = ([W - F + 2P]/S) + 1 \quad (2)$$

The output may be of $W_{out} \times W_{out} \times D_{out}$.

The convolution operation on image is represented in Fig. 5. Here, kernel, image, stride and activation map are separately represented. The next one is the polling layer which is used to reduce the spatial size of the image which is

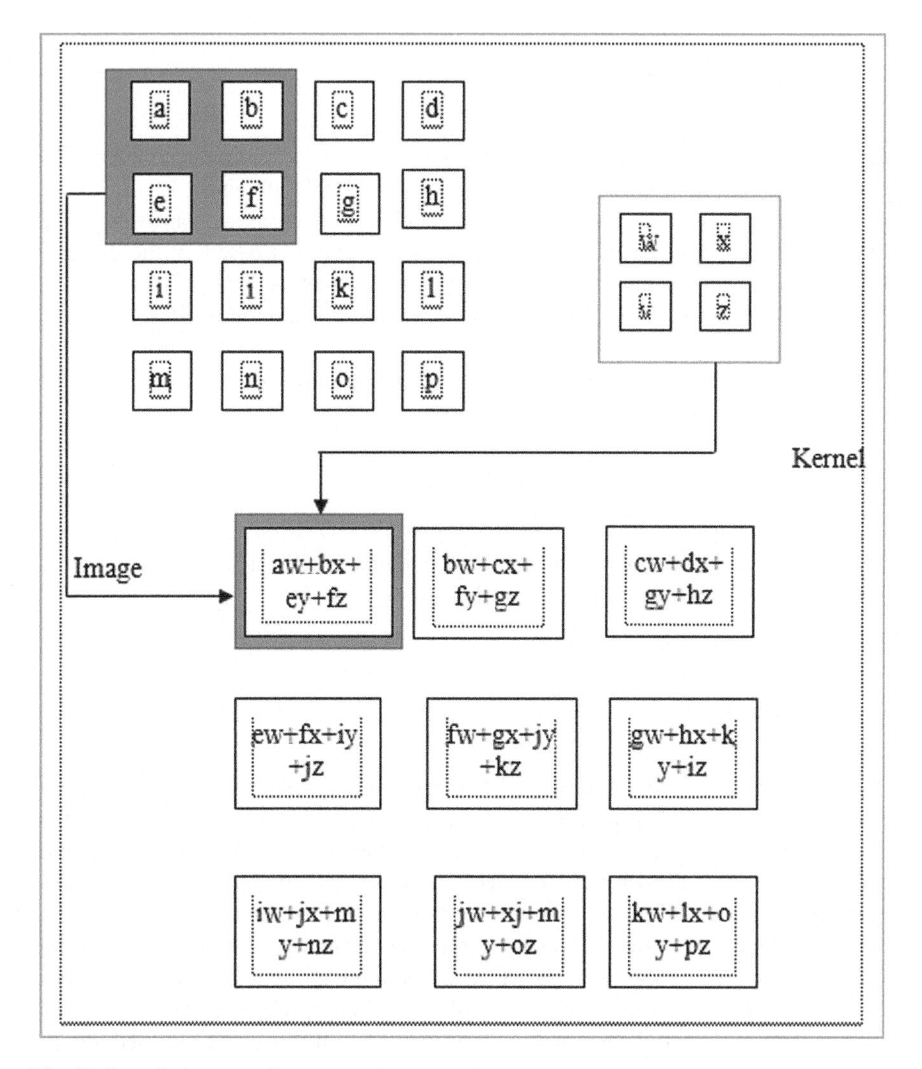

Fig. 5 Convolution operation

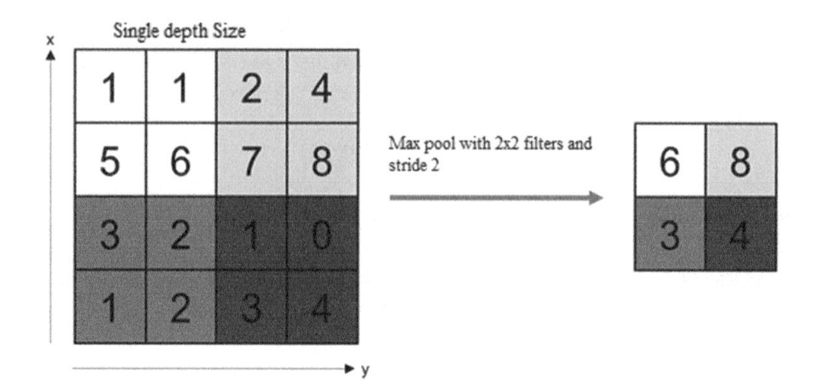

Fig. 6 Max pooling operation

represented in the binary values. Here, the summary statistic is done with the nearby output values. So that, the required size and weight can be obtained. It is made for every slides of the image. There are various pooling operations such as average, L2 norm and weight average of the rectangular neighbourhood based on the distance from the centre pixel. However, the most common method or operation is Max Pooling, which reports the maximum output from the neighbourhood.

The Max Pooling operation is explained in Fig. 6. If the activation map is of size $W \times W \times D$, pooling kernel of spatial size F, and stride S, then the size of the output can be,

$$W_{out} = ([W - F]/S) + 1 \tag{3}$$

It will the output volume of size $W_{out} \times W_{out} \times D$. In all cases, pooling provides some translation invariance which means that an object would be recognizable regardless of where it appears on the frame.

4 Methodology

The algorithm or the flow will be in the sequential manner. First, all the necessary software will be included for processing the images. So that, the in-built operations can be done by just referring the packages. Then the images or frames are acquired from the webcam in the form of the video. These are converted into frames and then taken for processing. They are then filtered to remove noises in the frame. Then the required segment is extracted. That is, required features are extracted from the frames for processing. [8, 9] proposed the different image segmentation techniques using Matlab and edge preserving for arieal colour image. These are then compared with the datasets for prediction and classification [10]. If it is not recognised, it will again be skipped to the previous phase. It is again pre-processed and taken for

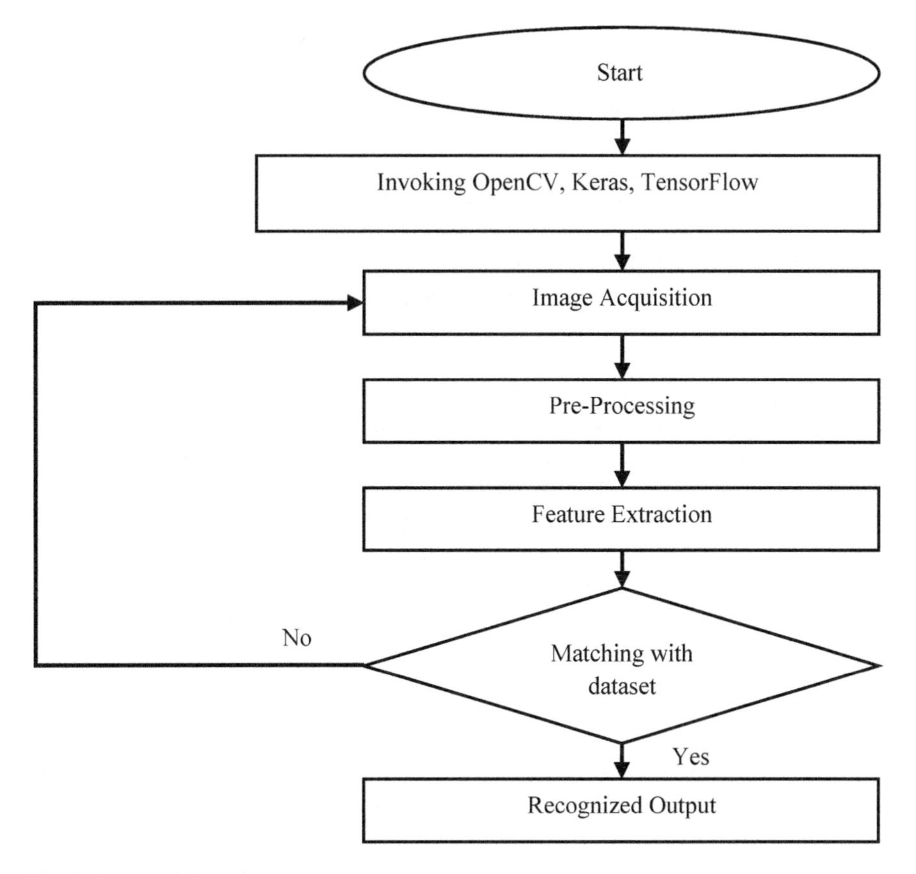

Fig. 7 Proposed flow chart

processing that is for recognition and classification. It is a loop which will occur repeatedly till it recognises an image. The methodology and flow of the process used for recognition and classification is shown in Fig. 7. This is basic one for recognition and after that, the image or frame will be compared with the dataset for prediction and classification [11]. First of all, many data are extracted and stored for Deep Learning process. [12, 13] proposed shape based image segmentation for hand gesture anaysis. From that, the output which is the classified image is determined the string is declared which is displayed on the screen.

5 Proposed Methodology

In this work, the gesture is recognized with the help of python. Here, first the input data (i.e. video) is obtained from the webcam of the PC. It is then snapshotted for Image Processing. Here, not all the images are captured. The coloured box is created at the right corner of the screen (i.e. Parallel to the Head). The things which are inside that box are captured and processed. Other places rather than the box are neglected. So, while the subject is inside the frame, his hand will be inside the box. It will be easy to process the image. First the images are resized, filtered and then converted into black and white. Then it is fed to the CNN. Here, all the required features are extracted from the image for training. Many images are collected for training. These results are stored for recognition. These are then tested for accuracy. The accuracy is very high when compared to other methods. Then, the recognition of live video takes place. Here, the live video is given as input. Same procedure is done for recognition. That is, snapshot is resized, filtered and proceeds to the classifier.

The block diagram of this project is shown in Fig. 8. The classifier will classify the image with the pretrained models. The result obtained will be displayed in another window. These results are used for many applications. It is a very useful methodology for the prediction of gestures. For example, if blind person is using the PC at home, He can control the whole system with the gestures. It can also be used for the Virtual Reality applications such as for gaming etc., Moreover it can also be used for controlling of robots, cars etc. In Fig. 8, the input is given as a video from the webcam and each frame is saved as image with same name. The image is taken for further processes such as filtering, feature extraction etc. In this, first, image is converted into black and white, then the noises are removed. These images are fed to the classifier. Here, the images are compared with the pre-trained models in the datasets.

First of all, all the images are trained and the results for the approach are stored. The tests also performed with the sample images for accuracy determination. After,

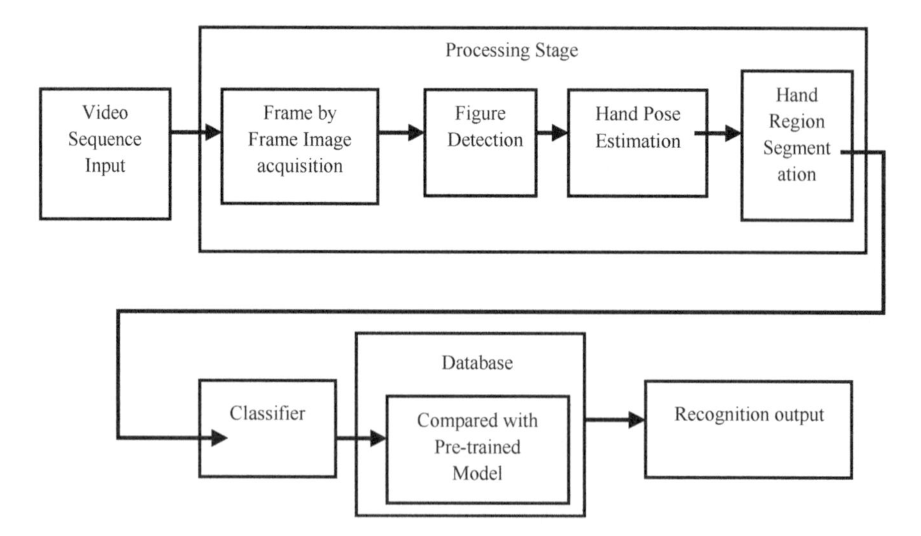

Fig. 8 Block diagram of proposed method

the live recognition takes place. Here, another window is created for the displaying of predicted results. The database is created with huge number of sample images for training.

The steps as follows.

Step-1 Data Obtaining—The data are obtained from the video which is obtained from the webcam. In the video only a particular region is selected for obtaining the images (i.e. frames from the video). From this, video is converted into 'n' number of frames which is determined by the frame rate specified.
Step-2 Data Pre-processing—Pre-processing can be done mainly with the help of two methods namely, Segmentation and Morphological Filtering. First method is the Segmentation [14], where the RGB image is converted into grey-scale image and then it is converted into binary image. Because, in the binary image only two regions are available. One is Background and another is Hand. The algorithm used is Thresholding technique. That is, values above the thresholding is changed into white and other is changed into black. So, it will be very easy to classify the images.
Step-3 Feature Extraction—In this stage, [15–17] pre-processed images are fed to the Neural Networks for training. In these networks, the required features are extracted and it will be matched with pre-trained models. Here, firstly many images are obtained for training. Above two methods are performed to remove noise and the features are extracted and saved. Huge number of sample images are processed for feature extraction. They are called datasets. They are used as the reference for recognition. Thus, the pre-processed image is compared and the matched one is considered as the output.

6 Software Used

The technologies which are used for [18–20] processing the image and software which can be supported for processing the images are listed below.

[A] OpenCV—It is abbreviated as Open Source Computer Vision Library. It is the library which contains many tools for real-time computer vision. It is an open source and free to download. It has C, C++, Java interfaces and supports many types of operating systems such as Windows, Linux, Mac OS etc. This library consists of more than 2500 optimised algorithms which also includes Machine Learning algorithms. From that, many complex algorithms can be performed just by calling the functions which has the pre-defined algorithms.

[B] Python—It is a very easy Programming Language which utilises Object Oriented Programming Concept. It is very special because, it consists of many libraries to perform many functions. We just need to include the library and we will be able to use all the functions in that library. It is a vast field which contains the packages of different domains. It is very famous because, it supports various package and many packages are also available in python.

[C] TensorFlow—It is the open source library for dataflow and differentiable programming across various places. It is mainly used to bring the Deep Learning and Machine Learning into the program. It consists of many functions related to Deep Learning and Machine Learning. That is, to determine the closeness of the features between the test and trained (present in the datasets).

[D] Keras—It is also another open source library mainly used for Neural Networks. It can be run along with TensorFlow. It has many functions which are related to the concepts such as CNN (Conventional Neural Network), RNN (Recurrent Neural Network), ReLU (Rectified Linear Unit) etc. It is very useful for Feature extraction with more accuracy (i.e. the more number of possibilities for success).

[E] NumPy—It is the important library which is used in python for complex mathematical calculations. These calculations mainly involve arrays, matrixes and for high level mathematical functions. These are very import because all the images are converted into matrix for processing and classifying. For all the libraries, this library will work along with it.

7 Experimental Results

Thus, by using methods like OpenCV, Keras, TensorFlow in python, the image is processed and features are extracted and by using DL method, the HGR is done successfully. The symbol and the corresponding alphabet is represented in Table 1.

The many samples are trained and many samples are tested. After that, the recognition also made and the results are discussed below.

The HGR is done and output is shown in Fig. 9. Here, the image inside the blue box will be processed and the results will be displayed adjacent to the test image. For testing, 'D' and 'M' gestures are taken and the images (or frames) from the video are processed and the processed image matched with the dataset and the corresponding alphabet is displayed on the screen. In the above example, D is displayed and M is displayed after it is processed.

In test cases, none of the gestures will also be placed on the box. So, another one variable is also made to be displayed and such a display is called 'None'. The test case also includes some unclassified gestures which is not in the dataset. That is, not a declared symbol or undefined gesture. For such a case also, 'None' variable is assigned. By this, 26 alphabets and a string, 'None' are the display parameters in this system.

The result shows none when nothing is placed inside the box is shown in Fig. 10. Here, the segmentation result is not matched with any of the images and the result is displayed as none. There is the algorithm which is to be executed or run

Table 1 Hand gestures and their corresponding alphabets

Alphabet	Hand gesture
A	
B	
C	
D	
E	
F	
G	
H	
I	

(continued)

Table 1 (continued)

Alphabet	Hand gesture
J	
K	
L	
M	
N	
O	
P	
Q	
R	

(continued)

Table 1 (continued)

Alphabet	Hand gesture
S	
T	
U	
V	
W	
X	
Y	
Z	
None	–

first which contains the methodology to extract all the features of the sample images. It is then tested with the test dataset for accuracy determination. That is, the error for the test cases.

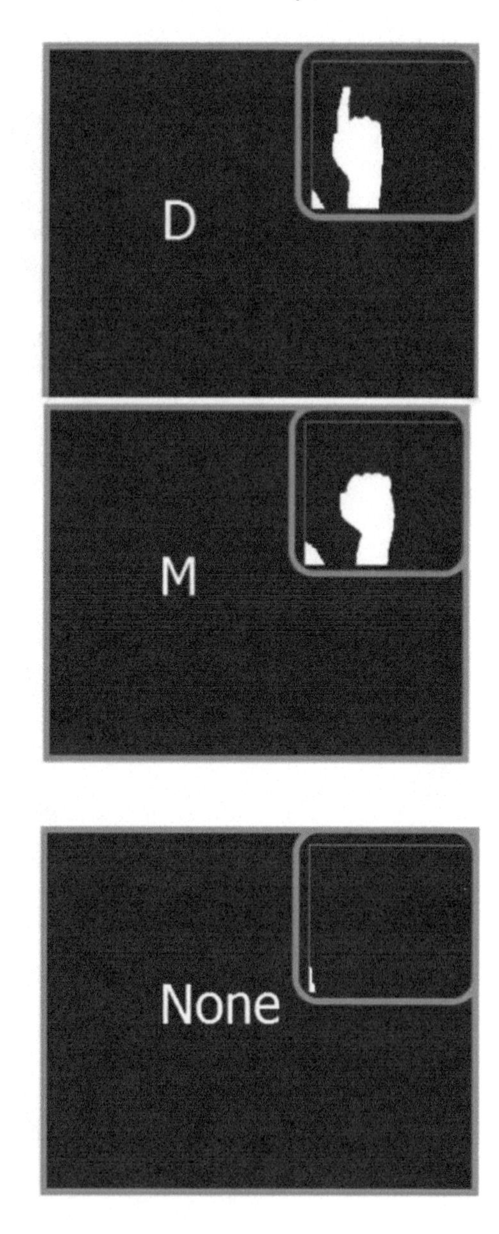

Fig. 9 Recognized hand gesture models

Fig. 10 Result when nothing is inside the box

Here, one graph is plotted between the accuracy and epoch and another graph is plotted between the loss and epoch. The epoch may be defined as the time period taken for completely pass the datasets in both forward and backward through the Neural Networks. So that, all the features are extracted effectively. In this project, each dataset contains 800 sample images. All these images will pass through the 2D CNN both in forward as well as in backward direction.

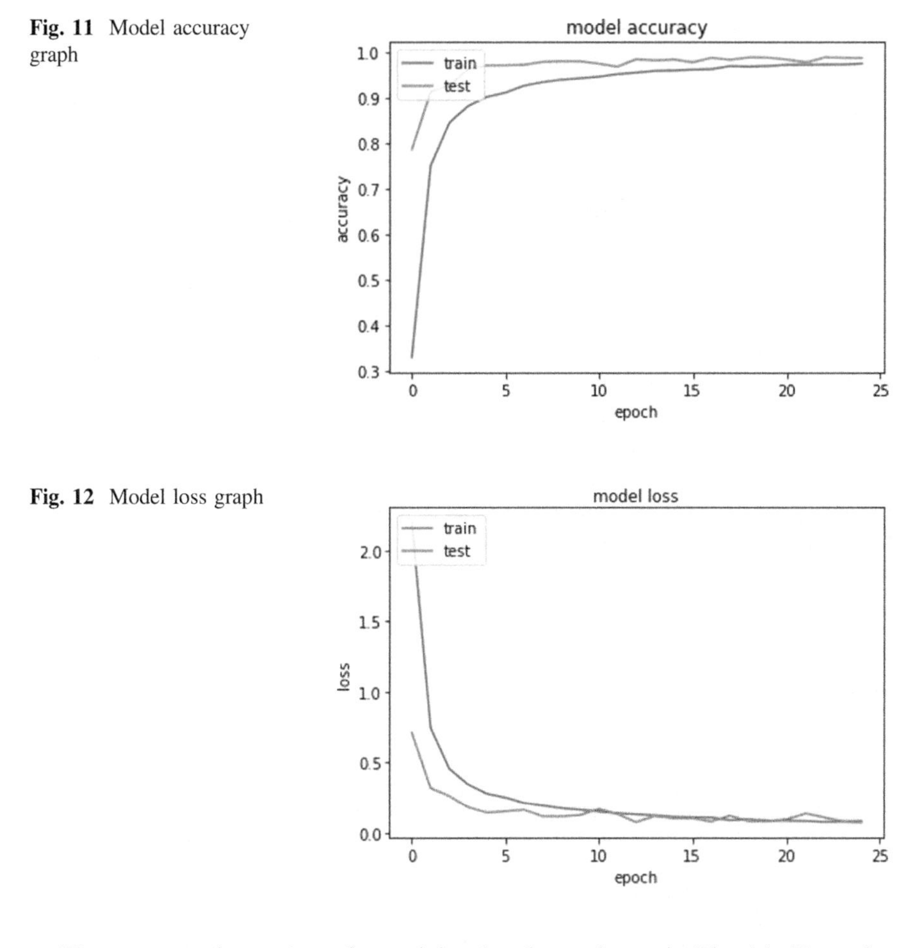

Fig. 11 Model accuracy graph

Fig. 12 Model loss graph

The accuracy for testing after training has been shown in Fig. 11. Here, the epoch is taken as x-axis and accuracy for y-axis. First, the accuracy slightly increases and settles down near one. After training, test is conducted and graph settles fast near one. That is accuracy increases after training. That is, first the training will occur and all the images in the dataset will be trained one by one and each one after training will be tested and it is called a single epoch. Likewise, nearly 800 epochs will be proceeded and after that, the testing will be done for error analysis. In this, for 800th epoch, it will be tested for the previous 799 dataset and the feature which differ from all the 799 images will be stored which is the huge process. After the training, the test dataset will test with the trained one and results will be obtained and this will be plotted in a graph for accuracy measurement. The accuracy is increasing after training. But, at first, accuracy is very low.

The loss for testing phase after training has been shown in Fig. 12. Here, the epoch is taken as x-axis and accuracy for y-axis. First, the loss slightly decreases and settles down near zero. After training, test is conducted and graph settles fast near zero. That is loses decreases after training. Here, loses means an error. The error occurrence is very high at first (i.e. before training) and it is decreasing slightly during training and it settles down. That is error percentage decreases after training.

8 Conclusion and Future Scope

On further development of this project, many datasets can be trained to increase the efficiency and the speed of the prediction also increased. It can also be used for Human Machine Interface, Virtual Reality Concepts etc. This gesture can also be used to control the laptops (i.e. without mouse and keyboard), control the robot, controlling of car. The interesting one is Bomb disabling robot controlled by the gesture. Here, the robot can be controlled by gestures from one side and it will work at another end. Not only a hand gestures, but also a face gestures also used based on the applications. That is, face gesture can also be used as a password for some times. In gaming, many virtual side will work with gestures which will be very realistic and interesting.

References

1. S. Shaikh, R. Gupta, I. Shaikh, J. Borade, Hand gesture recognition using OpenCV. Int. J. Adv. Res. Comput. Commun. Eng. **5**(3), 2275–2321 (2016)
2. M. Panwar, P.S. Mehra, Hand gesture recognition for human computer interaction, in *IEEE International Conference on Image Information Processing (ICIIP 2011)*, Waknaghat, India, Nov 2011
3. D.K. Ray, M. Soni, P. Johri, A. Gupta, Hand gesture recognition using python. Int. J. Future Revolut. Comput. Sci. Commun. Eng. **4**(6), 2459–5462 (2004)
4. A. Chitade, S. Katiyar, Color based image segmentation using K-means clustering. Int. J. Eng. Sci. Technol. **2**(10), 5319–5325 (2010)
5. Y. Guan, M. Zheng, Real-time 3D pointing gesture recognition for natural HCI, in *Proceedings of the World Congress on Intelligent Control and Automation*, China, 2008, pp. 2433–2436
6. E. Stergiopoulou, N. Papamarkos, A new technique for hand gesture recognition, in *IEEE—ICIP*, 2006, pp. 2657–2660
7. C. Herbon, K. Toninies, B. Stock, Detection and segmentation of clustered objects by using iterative classification, segmentation, and Gaussian mixture models and applications to wood log detection, in *German Conference on Pattern Recognition* (Springer International Publishing, 2014), pp. 354–364
8. D. Kumar, K. Kumar, Review on different techniques of image segmentation using MATLAB. Int. J. Sci. Eng. Technol. **5**(2) (2017)
9. B.N. Subudhi, I. Patwa, A. Ghosh, S.-B. Cho, Edge preserving region growing for aerial color image segmentation, in *Intelligent Computing, Communication and Devices* (Springer India, 2015), pp. 481–488
10. M. Panwar, Implementation of hand gesture recognition based on shape parameters, in *IEEE International Conference on Image Information Processing*, Dindigul, Tamil Nadu, India, Feb 2012

11. H. Renuka, B. Goutam, Hand gesture recognition system to control soft front panels. Int. J. Eng. Res. Technol. (2014)
12. J. Liu, W. Gui, Q. Chen, Z. Tang, C. Yang,"An unsupervised method for flotation froth image segmentation evaluation base on image gray-level distribution, in *23rd Chinese Control Conference* (IEEE, 2013), pp. 4018–4022
13. A. Jinda-Apiraksa, W. Pongstiensak, T. Kondo, A simple shape based approach to hand gesture recognition, in *IEEE International Conference on Electrical Engineering/Electronics Computer Telecommunications and Information Technology (ECTI-CON)*, Pathum Thani, Thailand, May 2010, pp. 851–855
14. A.B. Jmaa, W. Mahdi, A new approach for digit recognition based on hand gesture analysis. Int. J. Comput. Sci. Inf. Secur. (IJCSIS) **2**(2) (2009)
15. A. Sepehri, Y. Yacoob, L. Davis, Employing the hand as an interface device. J. Multimedia **1** (7), 18–29 (2006)
16. S. Phung, A. Bouzerdoum, D. Chai, Skin segmentation using color pixel classification: analysis and comparison. IEEE Trans. Pattern Anal. Mach. Intell. **27**(1), 148–154 (2005)
17. Z.Y. He, A new feature fusion method for handwritten character recognition based on 3D accelerometer. Front. Manuf. Des. Sci. **44**, 1583–1587 (2011)
18. Y. Zhao, C. Lian, X. Zhang, X. Sha, G. Shi, W.J. Li, Wireless IoT motion-recognition rings and a paper keyboard. IEEE Access **7**, 44514–44524 (2019)
19. T.H. Lee, H.J. Lee, Ambidextrous virtual keyboard design with finger gesture recognition, in *Proceedings of IEEE International Symposium on Circuits and Systems (ISCAS)*, May 2018, pp. 1–4
20. Y. Zhang, W. Yan, A. Narayanan, A virtual keyboard implementation based on finger recognition, in *Proceedings of International Conference on Image and Vision Computing New Zealand (IVCNZ)*, Dec 2017, pp. 1–6

Haptics: Prominence and Challenges

P. Sagaya Aurelia

Abstract Derived from a Greek word meaning "sense of touch", Haptic is a communication technology which applies tactile sensation for human-computer interaction with computers. Haptic technology, or haptics, is a tangible feedback technology that takes benefit of a user's sense of touch by applying forces, sensations, or motions to the user. These objects are used to methodically probe human haptic capabilities, which would be complex to achieve without them. This innovative research tool gives an understanding of how touch and its core functions work. The article will provide a detailed insight into the working principles, uniqueness of the technology, its advantages and disadvantages along with some of its devices and notable applications. Future challenges and opportunities in the field will also be addressed.

Keywords Haptic · Gloves · Tactile · Actuator · Haptic paddles · Haptic knobs · Novint Falcon · Force feedback gaming joysticks · SensAble's Omni Phantom

1 Introduction

Human-Computer interaction has today spread o a wide array of fields. Haptics technology has helped the user sense and acquires response through physical sensation as shown in Fig. 1. The five senses of human are taste, hearing, touch and smell and sight. Among these, the most proficient is touch. This sense is capable of providing simultaneous input as well as output. Humans can differentiate vibration up to 1 kHz via tactile sense.

VR/AR technology has become pretty good in augmenting and making a virtual object look real. Motion tracking technology and straight sensors can make our virtual objects replicate our real object. When various ways of sensing the world are

P. Sagaya Aurelia (✉)
CHRIST (Deemed to be University), Bangalore, India
e-mail: Sagaya.aurelia@christuniversity.in

D. J. Hemanth (ed.), *Human Behaviour Analysis Using Intelligent Systems*,
Learning and Analytics in Intelligent Systems 6,
https://doi.org/10.1007/978-3-030-35139-7_2

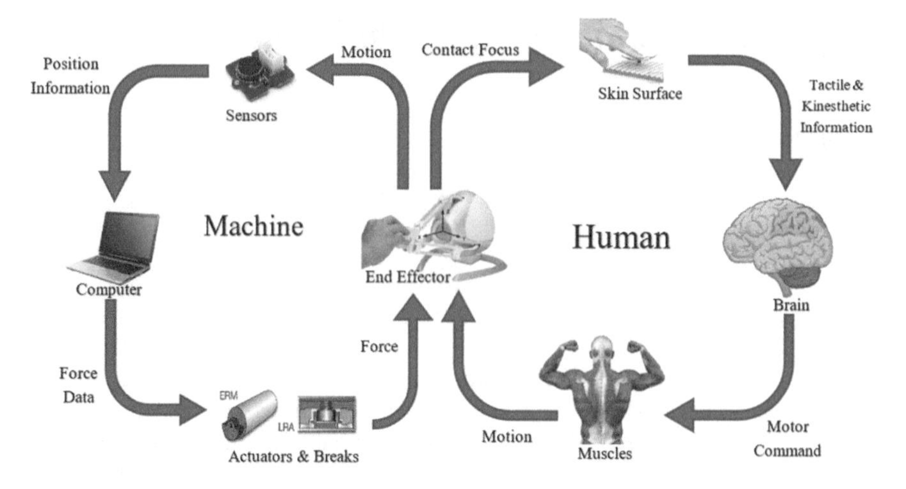

Fig. 1 Human machine amalgamation using haptics

considered and virtual reality replicates those experiences, humans' tactile senses is a vital factor in completing the illusion. This is where the field of haptics comes into play [1].

Haptic is derivative of a Greek word which has the meaning "to touch". Haptic is a combination of tactile information and kinesthetic information. Kinesthetic communication, also known as haptic communication, provides the user with the sense of touching by applying vibrations, forces or motions. Tactile content is the data derived from the sensor associated with the human body, while kinesthetic content is the data obtained by the joint sensors.

Haptic technology provides a touch feeling when vibration or any other force is given. Twenty-two types of touch sensations are produced by haptics to the user. The technology makes use of sensors, and based on movement sends an electrical signal. A computer in turns recognizes the signal and reverts a signal to the human with the help of joystick or wired gloves. It has analogous performances as that of what a computer graphics does for vision. Virtual object can be created and controlled using computer simulation. Haptic devices can calculate hasty and bulk forces produced by the user. As far as touch is concerned, a huge difference exists between cutaneous, kinesthetic and haptic.

2 History of Haptics

Researchers came up with all human touch-based experience and manipulation and the entire subfields are termed as haptics. Building a robotic hand and providing a touch-based perception and manipulation was a result of research efforts in the 1970s and 80s. It was during the early 1990s a fresh usage of the utterance haptics began to come into existence.

Today, haptics has a more significant influence on the growth of the automation industry. Researchers emphasized touch experience for human beings prior to the revolution in industries. Organisms, such as jellyfish and worms, in Biologists view, have a refined touch response. Medical scientists and researchers actively focused on how the human sense of touch functioned in the early 20th century. Human haptics, a branch of science, that looked at the sense of touch and feeling with hands was complicated to a large extent.

Researchers enumerate this handiness with the help of an idea known as degrees of freedom. The action or movement performed by a single joint is called a degree of freedom. Movements with 22 degrees of freedom are permitted as human hands contain 22 joints. The skin casing the hand is affluent with receptors and nerves, mechanism of the nervous system that interacts with the touch sensations to all the parts starting from the brain to the spinal cord. It is at that point of time where machine and robot's growth came into the picture. Then on researchers began to explore how these experiences can be sent to machines as the touch and feel of the environment was also experienced by the devices. That is the birth of machine haptics. Initially simple cable-controlled tongs were placed at the pole's end and the machines permitted haptic interaction with remote objects. Control tongs can be remotely handled, by moving, orienting and squeezing a pistol grip, which can be used to operate an object.

Manipulation systems were improved in the 1940s to provide nuclear and dangerous substance industries. Employees could handle toxic and hazardous substances, through a machine interface without risking their safety. Mechanical connections with motors and electronic signals were developed and designed by the researchers and scientists. This made it possible to communicate even subtle hand actions to a remote manipulator more efficiently than ever before. Remote manipulator became more efficient and it became likely to commune yet subtle hand actions. The electronic computer was the next advancement in the industry. In the beginning, the machines in the real world were controlled by computers. Later in the 1980s, the 3D world was generated by computers. During this period, the user was able to stimuli in terms of sound and sight only. Interaction with simulated objects with respect to haptic was within a boundary for many years.

In 1993, the Artificial Intelligence Laboratory of MIT came up with haptic stimulated devices and made it possible to experience the objects created by the computer. This domain was coined as computer haptic when researchers began to work on this project. Currently, a combination of hardware and software are usage to overlay the touch and sensation of virtual objects is termed bas computer haptics. Today, it is growing rapidly, providing a number of capable haptic technologies.

Various types of touch feelings in a haptic system are analyzed for future extension and usage. The first haptics patents go back to the 1960s and 1970s, out of places like Northrop Grumman and the legendary Bell Labs. Based on the earlier concept of multiple resource theory, the current haptics applications were developed. The kids were given an opportunity to experience and play virtual reality game and explore their controllers, which were made possible by tactical haptic.

3 Classification of Haptic

The most important obstacle for the growth of biomechanics and neuroscience is that the robotic stimulators are not able deliver a huge mixture of stimuli even though enough force and motion control are provided. User comfortability while wearing and using haptic interfaces is of dominant significance because ache and distress minimize all other feelings. Some examples of haptic interfaces are gloves and exoskeletons which are used for tracking and joysticks that can revert the forces to the user. Joystick which is attached to ground belongs to ground-based display and Exoskeletal is a body-based display which reflects force and is connected via user's forearm.

3.1 Wearable Haptics

Wearable body sensors and integration with computer systems are booming with the growth of Google Glass and Kinect. These devices entirely give a new view on human and computer coordination. Users from different background, mainly visually impaired who can get their distinguish needs be satisfied. In addition to this wearable haptics can also assist them are some of the other works such as driving in a group, food preparation, etc.

3.2 Touch Screens

Some of the main features of touch screens are a sound sensation, vibrations, attach and "experience" images on the screen. The steps and transaction involved from touch to host applications are shown in Fig. 2.

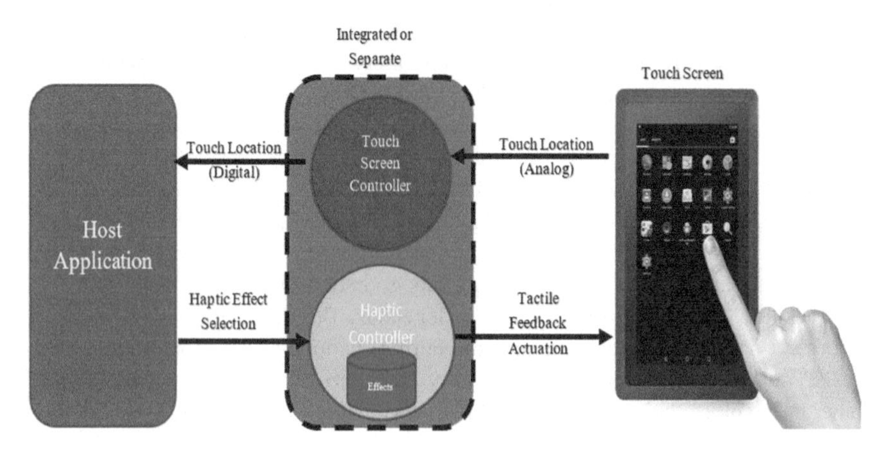

Fig. 2 Touch to the app—steps and transactions

A major categorization of haptic communications with real ambience or virtual realm that influences the interface is:

a. free movement, where no physical contact happens with objects in the atmosphere;
b. getting connected with forces which results in an unbalanced manner, for example, using a finger pad to press an object;
c. handling self-equilibrating forces and their connections, case of point, in a pinch grasp, when an object is squeezed.

Based on the usage and setup for which a haptic interface is designed, few or even all of them must be satisfactorily virtualized by the interface. While haptic interfaces are designed, the simulating elements are the major deciding factor for it. Accordingly, the interfaces can be divided based on the implementation of force-reflecting or not, category of motions (e.g. No of DOF) and simulating contact forces.

4 Haptic Devices and Interface

Haptic interface comes into existence where the user can receive instruction from the computer by the use of haptic devices, in a felt sense on a few fractions in human. For instance, in a virtual atmosphere, by making use of data glove, the user can pick up the ball. The system wits the action and throws the virtual ball there. On the other hand, due to haptic interface, the user shall experience the ball by tangible feelings that the data glove senses with the help of the computer.

As far as the hardware is concerned, ambiance with synthetic experience and haptic interface are still in their infancy stage. Most of the components available today are aggravated by the demand in predicting virtual ambiance. Straightforward position measuring systems are used for a long duration of time to offer control signals to the computer. They can be expressed in many ways, such as to engross or get in touch with the user without restricted efforts put on view and those that calculate faction without any contact. Application stirring growth of these devices lies between the range from the manager of gear to mech-biological lessons of individual movement.

Initially, developments in haptic interfaces for force-displaying were determined by the requirements of the nuclear energy and others for isolated exploitation of resources. The credit of the need for excellent strength display by untimely researchers continues to be pertinent to today's VE application. VE and haptic media interface are shown in Fig. 3. Even though the ground-breaking explanation of VEs incorporated interfaces with force-reflecting, expansion of realistic components has demonstrated to be complicated. An uneven breakdown of the foremost type of haptic interfaces that are at this time obtainable or being urbanized in laboratories in different regions of the world are shown in the following picture.

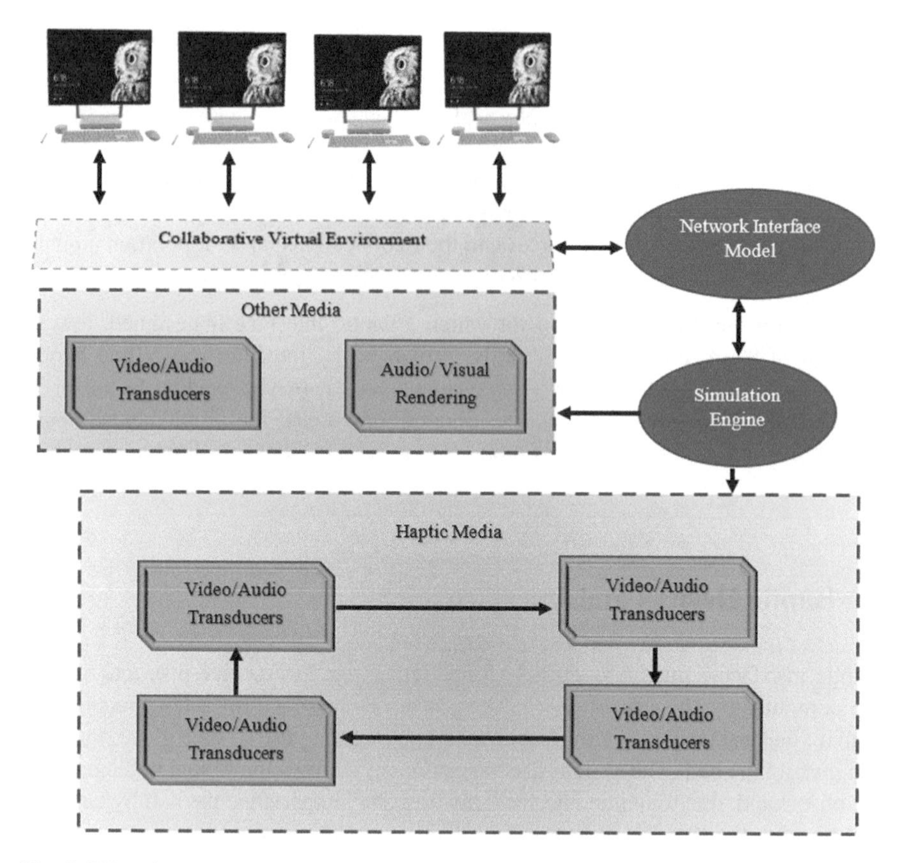

Fig. 3 VE and haptic media interface

The various types of haptic devices such as ground-based, body-based and tactile display are as shown in Fig. 4. Joysticks are almost certainly the oldest of technologies and were initially utilized in command of aircraft. Even the most basic of control sticks, associated by habitual wires to the voyage surfaces of the aircraft, unintentionally obtainable force information. In broad-spectrum, they may be inert, as that of joysticks worn for cursor position, or lively as in today's contemporary flight-control sticks. For instance, Measurement Systems Inc. has sold quite a lot of 2- and 3-DOF joysticks, a number of which can intellect but not exhibit force. Instances of force-reflecting joysticks deliberate for comparatively greater BW are the AT&T mini-joystick [2–4].

Numerous hand controllers which reflect force, obtainable nowadays are urbanized to be in charge of remote manipulators. By and large, the components utilize at most six degrees of freedom and have extensively higher recital individuality. Predominantly high-quality review of recital uniqueness is found and a great impression of the components came to usage by Honeywell. An immense

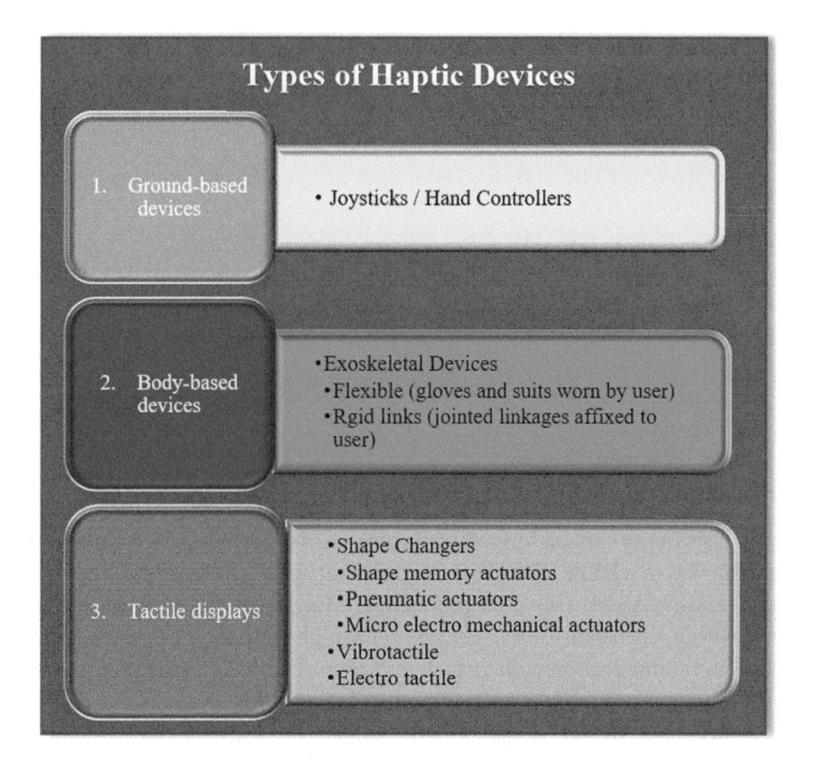

Fig. 4 Types of haptic devices

effort relating to ergonometric (form, switch position, movement and power distinctiveness, etc.) occurs due to the hand grip of these instruments [5, 6].

4.1 Phantom Interface

a. Open and transparent.
b. It provides a 3D touch of the virtual object.
c. When the finger is moved by the user, a virtual 3D object program can be really felt.

4.2 Cyber Grasp System

- The cyber grasp system is a pioneering force feedback system for user's fingers and hand.

- When a force of feedback system with innovation is used by the user's finger and hand then cyber grasp system comes into the picture. Magnetic levitation haptic interface for total body sensation, this type of magnetic levitation haptic interface is used.

Benefit

- Operational time is reduced.
- Interaction is centralized through touch so that the digital realm can imitate like the real world.
- In gamification, medical and in military it is used wide apart.

Drawback

- Higher expenditure.
- Huge heaviness and volume.
- The precision of touch needs a bunch of progressed design.

One of the primary applications of hand controllers reflecting the force to the virtual environment was in scheme GROPE at the University of North Carolina. The Argonne Mechanical Arm (ARM) was used productively for force reflection when communications within. In recent times, devices with high performances are exclusively intended for communication with virtual ambience. The MIT Sandpaper is a 3-DOF joystick that is competent of virtual textures display. PER-Force is a 6-DOF hand controller provides greater response and recital. The, built in the MIT Artificial Intelligence Laboratory introduced PHANTOM, which is a low-inertia device providing multiple links that can give a sense of virtual objects. Another high-performance force-reflecting master is a system operating on a group to facilitate two hand teleoperation of a microrobot that can convene the twofold necessities of bandwidth and elevated accuracy. An enhanced version is used for eye surgery based on teleoperation and signifies that it can be obtained using current technology.

Exoskeletal devices are identified as they are intended to be fit over and move the fingers of the consumer. Because they are kinematically comparable to the hands they check and arouse, they have the benefit of the widest variety of unhindered user action. Exoskeletal devices which are used to measure position are somewhat low-priced and at ease to use.

The renowned VPL Data Glove and Data Suit use fibre optic sensors to attain a shared angle motion more than a degree. The Virtex Cyber Glove is able to get a higher resolution for half a degree with the help of strain gauges. Rigid exoskeletons devices that offer force reflection in adding together to joint angle sensing are also built. Conversely, as long as high-quality feedback, with devices like that which is proportionate with the individual resolution is complicated and results in enormous difficulty on reducing the size of the actuator. While the demonstration of net forces is suitable for imaginary, virtual object communication, researchers have also familiar with the necessity for more accurate displays surrounded by the contact region. In fastidious, tactile information are displayed, has long been measured enviable for inaccessible handling.

Tactile display system within the last 20 years has been mostly used in transmission visual and audio content to deaf and blind persons. Display systems that take effort to put across contact data about contact apply a diversity of techniques. Displays which changes shapes put into words the local shape by calculating the forces across on the skin. This becomes possible by many stimulators actuated by DC solenoids, form memory alloys, and dense air. The employ of a permanent surface actuated by an electro rheological fluid has also proposed. Tactile with vibration displays transmit mechanical energy throughout multiple vibrating pins placed alongside the skin. Electrotactile shows kindle the skin from beginning to end surface electrodes. An assessment of main beliefs and technological issues in vibrotactile and electrotactile displays are also found. The virtual substance with physical model and ambience take delivery of user's instructions all the way through the haptic interface sensors and produce outputs analogous to the corporeal performance of a virtual entity in the virtual world.

5 Haptic Feedback

There are four diverse types of hardware most repeatedly used to afford haptic feedback:

5.1 Eccentric Rotating Mass (ERM)

Eccentric rotating mass (ERM) actuator being the most popular type of haptic feedback based on vibration is an electric motor which rotates with a mass. The force of the offset mass is asymmetric as the ERM rotates. These results in a net centrifugal force, which makes the motor visible. As it swiftly spins, the motor is continually displaced, which introduces the emotion for vibration. The possessions that the user receives from ERMs are low.

5.2 Linear Resonant Actuator (LRA)

The subsequent increase in haptic modernism is the linear resonant actuator (LRA). It comprises of a magnet close to a spring, bounded by a coil. It's determined by the coil that is animated. The mass moves within the coil, which forms vibration.

LRAs became trendy and are at a standstill found in the GS8 Smartphone and [–].

LRAs are also connected to ERMs because they use limited power and have a quicker retort time, making them available simulated usage for texting apps on mobile phones.

Nonetheless, this enhanced haptic presentation over ERMs comes with a less component cost. Identically, competence and performance reduce significantly as the LRA's drive occurrence moves exterior of its resonant group.

5.3 Piezoelectric Actuators

Piezoelectric actuators consist of ceramic material to generate motion and force. Piezoelectric actuators are further accurate than together ERM motors and LRA's due to their capability to vibrate at an extensive mixture of frequencies and amplitudes which are separately managed using AC voltage. In view of the fact that the vibration doesn't trust on the resonant frequency of spring, but in turn, it may be customized liberally excluding defeat of efficiency. Kindle Voyage is an illustration of a device that has within itself piezoelectric actuators. The three criteria determining the higher level of piezoelectric actuators are the component's rate, delicate characteristics of the materials, and power expenditure, as it is in need of voltages more than ERM and LRA.

5.4 Accelerated RAM

The forced impact also known as accelerated RAM, contains a magnetic hammer balanced by air. TacHammer one of the latest releases based on accelerated RAM. This is found in TacHammer the company which is associated with the nanoport. It's able to produce similar outputs, including hard tap and soft tap. This is also utilized to simulate emotions like heartbeat. The hard tap is introduced by using the hammer in the reverse track, producing an augmented RAM striking, affecting the realistic gunshot, outcome of the kick. The TacHammer uses noticeably a very minimal power usage than the LRA and ERM, and there can be variations in voltage and the frequency, in which the developers are provided the broadest spectrum of special effects to make use of, without reducing the real power.

Haptics or haptic feedback technologies are categorized into five different types such as ultrasound, thermal, force, vibrotactile or electrotactile.

High-frequency sound waves are ultrasound. Ultrasound feedbacks are created using either single or multiple emitters. The emitter from one part sends information in the form of signal to another part of the body which is called as "acoustic time-reversal", the principle of transmission. This ensures that for larger areas, haptic feedback field is a must. This is not possible using a single emitter, but the enormous number of emitters are appointed for this task. An invisible tangible interface is created altogether in the air. In the air, the user can feel the turbulence generated by the ultrasound waves. Usage of any additional accessories is not a must in ultrasound technology, which is the main advantage. On the other hand, high cost and less perceptible are some of the notable disadvantages.

Actuators' grids are used in forming thermal feedback. It has a direct touch with the skin. People don't define well the place of the thermal stimulus are not well defined when comparing to tactile interaction. In turn, a very a smaller number of actuators can be used and also the position need not necessary be to close. For these reasons, this type of feedback is very easy to design. But still, heat cannot be taken from many places due to the law of energy of conservation. Only the movement from one place to another place is made possible. Beyond this to give a realistic experience, this should happen very rapidly. Henceforth a lot of energy is used in these types of haptic using thermal feedback.

Haptics uses vibrotactile feedback most commonly. To specify human skin receptors these stimulators, apply a huge amount of pressure which can vibrate up to 1000 Hz. Since the frequency range from 80 to 250 Hz is the limit for human speech frequency, due to which the skin feels the sound. There are various merits such as cheap, simple and it can be easily powered and controlled. Since for a long time this type of feedback is used in devices like mobile phones and steering wheels. In addition to this, it is need of only very less tracking parameters and power consumption. On contrary to this, in vibrotactile feedback also has few drawbacks such as depth and diversity of sensations cannot be given by vibrating motors. Vibrating motors are difficult to miniaturize competently. Some of notable medical side effects of vibrotactile feedback are causing negatively affecting ligaments and joints due to deep penetration, moreover, ghosting effect also appears with extended or powerful impact.

When wearable haptic suit is used, the user can feel the breeze or texture. This becomes attainable due to electrotactile stimulators based on electrical impulses. Any sensation can be imitation with electrical impulses. Based on the intensity and frequency of the stimulus, various subtypes of electrotactile feedback can be formed.

Factors such as electrode size, hydration and skin type, contact force, voltage, material, current and form of a wave the sensation in this type of feedback can vary. It is due to electrotactile feedback a diverse form is achievable, due to its higher potential. Absence of mechanical or moving parts is one of the greatest advantages over the other type of feedback. The user obtains high sensations due to electrical impulses which is not possible in other currently exist feedback systems. Customization of electrodes into array is the foremost a merit of electro-neural stimulation. Electrical muscle stimulation is used in medicine for 30 years and has given good results in safety. Since electrical signals are the core of the nervous system, Electro-tactile feedback is more applicable for creating and simulating sensation.

Force feedback being the first type of feedback every introduced, it affects the muscles and the ligament via our skin keen on the musculoskeletal system. There are two kinds of force feedback devices such as, biomimetic and non-biomimetic. The movement of biomimetic devices is based on human limbs. Such devices are difficult to develop. It is very difficult to develop such devices due to the requirement of having a functionality of the human body and on the other hand suitable for various people. When an unusual model from the human body is needed then non-biomimetic devices are used.

Based on appliances, force feedback appliances are classified into resistive and active one. The movement of the user is limited by brakes in resistive devices whereas, in active devices, a movement of the user is limited by motors. In addition to this, active devices can also simulate various interactions, on the other hand, they are also much more complicated to control.

6 Software and Inference of Haptics

Universally, haptic interfaces accept commands for motor action from the user and in turn as an output it displays relevant pictures to the user as shown in Fig. 5. Tactile pictures contain force as well as displacement fields which will be overlaid on the viewer to trigger the user's wished mechanical communication with 3D objects in the virtual ambience. Most commonly, these images trigger tactile as well as kinesthetic data medium and are executed by the actions. Foremost instruments of the data delivered are connecting modes with the objects, mechanical properties of the objects, and the motions and forces associated in investigation and handling in a virtual ambience.

Since haptic interfaces for communicating with virtual ambience are in the infancy stage of expansion. The software that is predominantly designed for creating tactual images is extremely limited. Codes compulsory used for position tracking are also offered for commercial purpose. On the other hand, as for as force-reflecting devices are concerned, as hardware, many software has been introduced in teleoperation view or calculating autonomous robots. Quite a lot of research laboratories

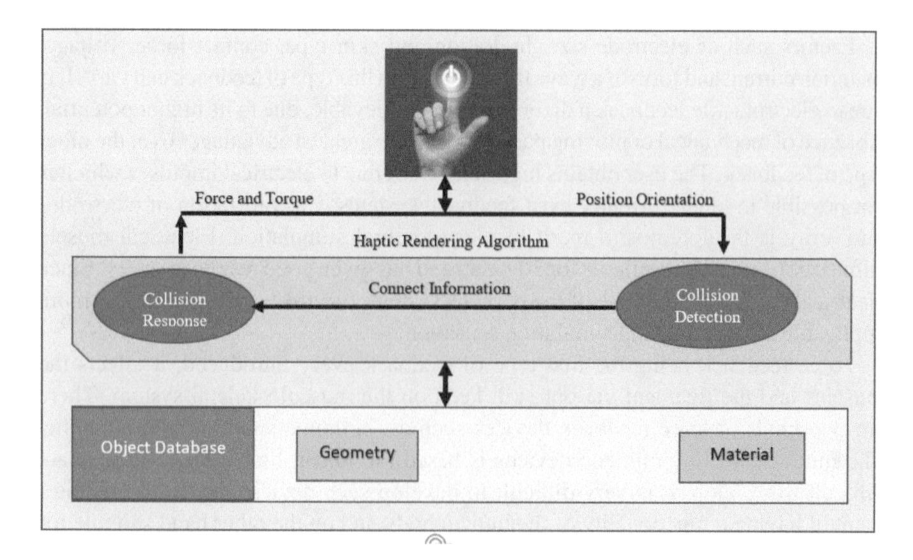

Fig. 5 Haptic software

have developed based on virtual ambience systems with visual and haptic output attainable via suitable amalgamation of virtual objects and interface for rendering tactical images with software used to make use of visual images.

Parallel to the software desired to create visual images, the software desired to created tactual images can be divided into three groups: the virtual object's model and surroundings, rendering software and interaction software. Haptic interaction software primarily consists of understanding the interface device. For instance, the signal taming and noise lessening software essential for interpreting sensors of this type. Instruments used for analyzing hand posture, a software relying on the kinematic model of the hand is required as in case of exoskeletal, and also for examining the signals as parallel to the position of the hand.

Assessment of haptic interfaces is critical to evaluate their usefulness and to separate features that require development. Nonetheless, such estimation performed in the perspective of teleoperation has been so mission-oriented, so that it is unfeasible to get helpful uniformity and to outline efficient hypothetical **models** relied on the normalization. There is also a need to identify a set of basic physical tasks that can assess and contrast the physical abilities.

7 Working Principle of Haptic Technology

Haptic sensors are already into existing ever now and then in different forms. Regardless of the type of haptic technology utilized, irrespective of the type of usage of haptic technology the underlying working principle such as motion, force and vibration and their combination in creating a touch experience is the same. Haptic has a lot of advantages such as enlarged user approval and more practical understanding. Furthermore, it also increases the task performance. The working of haptic technology is shown in Fig. 6. The essential standard of a haptic sensor is the

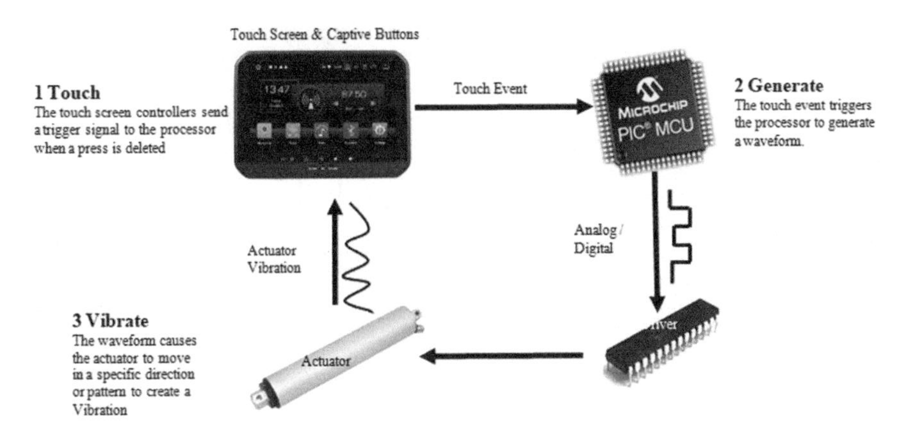

Fig. 6 Working of haptic technology

production of an electric current that results a reaction to generate a vibration. The possibility of achieving this is where the diverse technologies be at variance.

Nonetheless, not all haptic sensors are in need of touch to work. They are called as non-contact haptic and make use of ultrasound technologies to generate 3D space in the region of the user. The user in turn communicates with the space just about a device excluding the want to actually touch it [7].

Proprioceptors are the two types of feedback the tactile and kinesthetic which are received when we use our hands to explore the world. These receptors are responsible for passing signals to the brain the somatosensory region processes the information.

The information on changes happening in muscle length is given by muscle spindle (a type of proprioceptor) whereas Golgi tendon organ provides information on the changes happening due to muscle tension. This kinesthetic information is processed by the brain to afford a sense of the object and shape and position connected to the body, or hand and arm. Different receptors are used for various stimulus operations such as pressure, vibration, heavy touch, pain and light touch. These various data from receptors assist the brain in understanding various tactile information about the object.

There are quite a few appeals in a haptic systems creation. Two significant things, in general, are a device which applies forces and are connecting with the user and software to find the output forces when a user's interconnects with the virtual object. The software process used for calculation performance is called haptic rendering. The steps involved in rendering haptic data are shown in Fig. 7. A widespread rendering method uses polyhedral models to signify virtual objects. These 3-D models can precisely represent a range of shapes and evaluate the force line's interaction with different faces of the object which in turn determines the touch data. Such 3-D objects put forward a solid experience and also gives the texture of the surface. The interface device delivers the haptic images to the user.

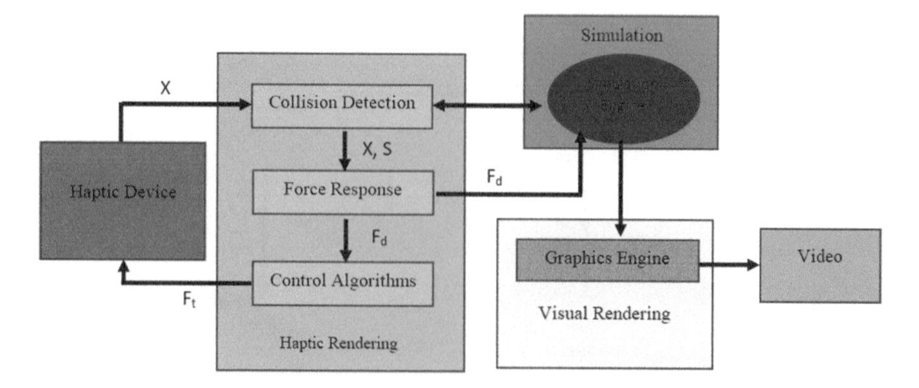

Fig. 7 Haptic rendering

8 Factors Determining Haptic Actuator or Haptic Sensor

Various factors should be considered while selecting haptic sensor or haptic actuator. Some of the four major factors are as follows:

- **Escalating of haptic device:** It influences the genesis of vibrations and how user understands the possessions.
- **Retort time:** The time rate in producing certain effect. (The fastest response)
- **Power utilization:** It is based on vibration period and mass in moving state.
- **Bandwidth:** Frequency series of possessions it can create or spawn.

8.1 Input-Output Variables of Haptic Interactions

In virtual world, the haptic interfaces obtain the proposed motor action commands and delivers tactual images as output to the human. Displacements and forces, their spatial and temporal variance are the primary input-output variables of the interfaces. Henceforth a correlation between forces and displacements in diverse points and orientations on the skin surface generates mechanical impedance in haptic interface. Depending on the control algorithms used, either displacement or force are control variable, and the other is a display variable Accurateness is exaggerated by pointing a finger, direction, the speed, and magnitude of movement in locating a target position. Examining the human ability in calculating limb motions has classically deliberated human tracking performance. The thresholds for position and movement are 8%. Human BW for limb motions works in 1–2 Hz for unanticipated signals and 2–5 Hz for cyclic signals, up to 5 Hz for trajectories generated within, and >10 Hz for reflex actions.

8.2 Overall Force of Contact

Whenever an object is pressed, both the tactile and kinesthetic sensory systems sense the active motion of the hand and the contact forces. The neural signals and the control of contact in the sensory system situation are decided by overall contact force. The greatest controllable force is exerted by a finger pad is regarding 100 N and the resolution in visually tracking steady forces is 0.04 N or 1%, whichever is elevated.

8.3 Insight of Contact Circumstances and Object Characteristics

For minimalism, standard notch, lateral skin stretches, qualified, and vibration are the conditions of contact primitives of object. Primitives of object properties are obtained by touch based on various features such as Surface micro texture, shape and compliance. Cutaneous mechanoreceptor delivers tactile information in human perception. Neural codes are future divided as intensive, temporal, spatial, and spatiotemporal.

9 Haptic Scenario, Market, Future and Applications

Motion, vibration and force and their combination results in touch experience which is recreated by haptic sensors or haptic interface devices. Irrespective of the industry be it be game console controller or smart phone or automobile haptic technologies are growing very rapidly and used everywhere. Just by manufacturing and implementing haptic sensor will reach a maximum revenue of $12.8 billion. Haptic technology market, throughout the world deals with current market valuation, dynamics segmented by various region such as region, country level usage, share analysis and supply chain analysis and also markets trends in the past.

Enhancement in usage of electronic devices such as mobile phones, tablets, automotive sectors and gaming consoles is one of the major reasons for the growth of haptic technology market. The device which measures the force released by the user is connected along with the haptic devices. The user experiences a novel realm with the help of haptic user interface.

The various factors determining the market of haptic technology are feedback, application and components. Haptics market by components and applications is shown in Fig. 8. Future more the component is divided into driver, controllers, software and actuators. The software component market will grow maximum among all the other elements.

This is majorly due to the applications in various sectors such as health care, gaming, automotive so on and so forth. Over 16% of CAGR between 2018 and 2023 and growth of approximately USD 22 billion by 2023, is the growth rate prediction of universal technology market. Haptic market by components and by applications from 2016 up to 2024 is shown in Fig. 9.

The provincial investigation of haptic technology market was done for area such as Asia Pacific, North America, Europe and Rest of the World. The investigation results states that the largest share of market arises from Asia Pacific as shown in Fig. 10. Technical growth and rising insist for mobiles and tablets are also expected in Asia Pacific region. Whereas, within the forecast periods North America shows a fastest rate.

Fig. 8 Haptics market by components and applications

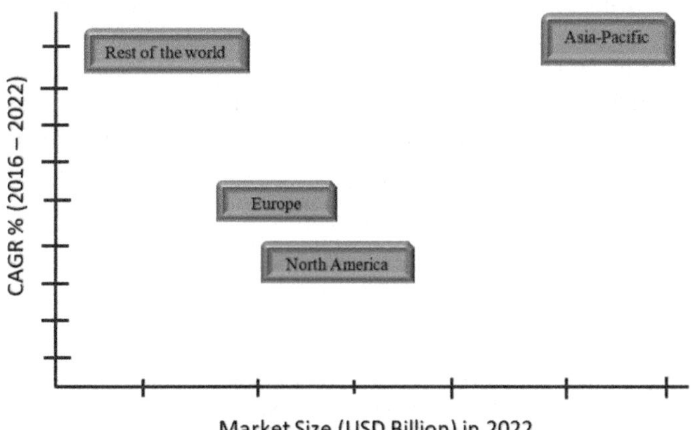

Fig. 9 Region wise prediction on haptic market

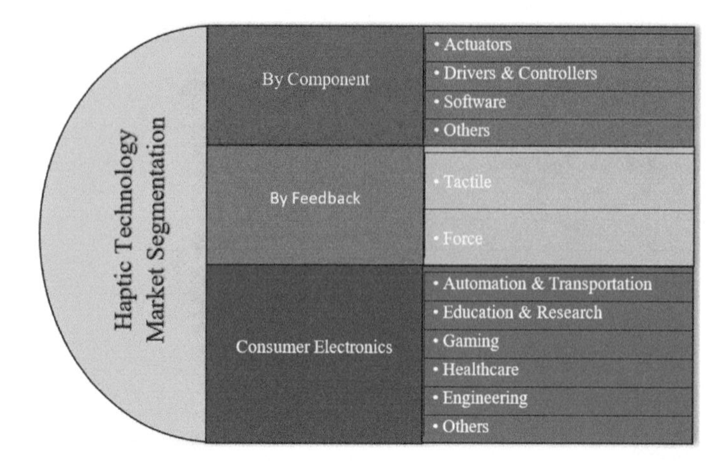

Fig. 10 Haptic technology market segmentation

9.1 *Haptic Technology Market Goals*

- To offer forecasting of a mixture of segments and sub-segments of the haptic technology market as shown in Fig. 10 and comprehensive analysis of the market organization.
- To supply understanding on criteria influencing the market development.
- To afford country-based study of the market focusing on the current size of the market and future perspective.
- To make available country wise investigation of the market for segmentation on the root of component, response and function.
- To endow with planned profiling of key players, broadly finding their main competencies through which a aggressive background for haptic technology market can be laid on.
- To track and analyze cutthroat growth such as collective business planned alliances, novel item growth, investigations and advancements in the market related to haptic technology.

9.2 *Applications*

- Haptic technology is a powerful tool which can be used in all fields where constant investigation of human Haptic capabilities are required. It provides human sense of touch. The technology can provide some remarkable outcomes in enormous sectors and it can be future extended to advanced level. Haptic technology provides an open platform for various related potential fields also.

A servomechanism system based modern aircraft was the first ever Haptic device-based application. It was basically used to operate control systems.

- Haptic is extensively becoming a device used in a large quantity of applications in the healthcare trade. Sense of touch is decisive for medicinal guidance. It can be used to assist the visually impaired by allowing users to touch a exhibit and also experience the texture on their fingertips. Other uses could be in medical guidance with simulators which uses virtual reality, implanting come within reach of that bring into being criticism or even in robotic surgeries. In medical industry, it must directly be likely as a doctor looking at a 3D MRI or CT image and further more sticking his hand into the image and experience for things like blood clot, blockages and tumors [8]. Case in point, medical trainees can now perfect subtle surgical strategies with the help of computer, respond to stitch blood vessels in an anastomosis. In future, an assortment of haptic interfaces for therapeutic simulation may confirm in particular constructive for instruction. Doctors doing remote surgery through robotics will need to have active response of force and movement [9–11].
- Teaching with haptics is becoming more and more widespread. Newly, Rice University urbanized a haptic rocker that permits users to grasp bits and pieces with a hand of prosthetic type to confer a muscle and artificial limbs sensation. Thus, the technology impact could be used to alter how traditional prostheses purpose and how amputees can grasp objects. A non-invasive edge with rotating arm is used above the skin by the expertise user to brush the rubber pad for feedback. In edification, haptics could be used as a way to tutor students or develop the learning familiarity for students, in particular those that are falling last or those that have learning deficiency [12].
- Haptics in cars is becoming a way for automotive displays to afford tactile feedback for superior control inside the car and also provides better sense of safety. Force feedback has been used beforehand for automotive industrialized simulation applications, more than ever for assembling and developed components. At the present, it is being ported surrounded by the vehicle for applications such as in receipt of response from the steering seats as a vibration demonstrating a pedestrian is about to cross the street. Bosch came up with a concept car with haptics technology integrated. Ultra-haptics at CES 2017, Bosch showcased haptic feedback for gesture control surrounded by cars to control diverse functions inside the car. Ultrasound wave's touches the user's hand and endow with reaction that makes it undergo as if the user is in fact moving a physical knob [13].
- Robots maneuver the surroundings by relaying data to a essential computer for analyzing and processing the data. Haptic technology could significantly support elderly people with disabilities and limitations. Haptics is by now growing rapidly in robotic industry. The surgeon, hold the control and will experience what the robot is sensing and in turn act in response appropriately. In robotics, haptics can also be used when someone is trying to revamp a piece of equipment distantly. Having some kind of oblige feedback ability would accelerate the service procedure. Or else when a robot is sent to a hazardous or inaccessible

vicinity and its machinist needs feedback to make certain that the machine has proficient in its duty. The information is feed by the airbus pilot and the airbus is in turn responding. The pilot gets the information when he is approaching a stall and, optimistically, responding [14].

- Computer researchers in Greece are using it for visually impaired to use touchable maps. Software examines the video to decide the figure and position of every object. A haptic interface device is used; visually impaired people can sense these forces with audio cues and get a enhanced feel of building's layout. The method can exert with Braille, a blind person can feel by altering every non-Braille into Braille characters [15, 16].
- Working with aircraft mechanics, multifaceted parts and examine actions, emotive the whole thing which is viewed on the monitor. And soldiers are prepared for combat in a assortment of ways, from learning how to determine a #bomb to work a helicopter, fighter jet, or in any virtual combat atmosphere [17, 18].
- Haptic technology is in advance esteem in the mobile user technology field, in order to provide vibration feedback on smart phone touch display. Material Haptic feedback is becoming fundamental in cell gadget. A handset maker as well as typical sorts of haptic novelty in their gadgets takes the materialization to feel the vibration response. Electronics utilize a haptic technology called plus touch for touch-screen auto route units [19, 20].
- Remote-controlled robotic tools that uses human operators to control isolated or far-off ambiance also utilizes haptic technology. The actions of the robots are controlled by human, in a telerobotic system, that is positioned at few meters far way. A few telerobots are restricted to extremely easy tasks, such as focusing a camera and responding images with visual effects. In telepresence which is an extended form of teleoperation, a sense of being is being sensed by the human operator within the ambience of the robot. In telepresence model, due to the potential of haptic, it takes into account the touch points and in toiling both the video and audio information [14, 21].
- In gaming, amalgamated reality will join with wearable technology to generate novel gaming vests that give response as a reply to what is occurring in the interior realm of googles. As vibration in a mobile phone indicates another call, vibrant searching makes a "haptic dialect", deciphered likewise altering the movement of the sign. Tactical Haptic, has handheld devices for games that replicate inertia and weight. The sword's weight, will be sensed when it is lifted by the player in a combat game, swings it and monitors in the display, the movement of the sway and the collision are felt. To a large extent of the prospect knowledge will come from game progress.
- Designers makes use of suppleness in a larger level data gadget that provides touch feedback examining the "surface" they are creating, allowing faster and greater standard work procedure than habitual techniques.
- Video game creators have been near the beginning of inert haptics adoption, which uses vibrating joysticks and steering wheels focuses on display. However new video games provide an opportunity for the players to undergo and experience virtual solids and tools. OS such as MaC and Windows will reach a

greater profit margin because of haptic communications. Touch screen Companies are already experimenting of providing a graphical button feeling for graphic buttons with this technology. Nokia introduced a tangible touch screen that makes replacing real buttons with on-screen buttons. When a button is pressed, movement in and movement out feeling should be attained. An audible click can also be heard. It was already accomplished by Nokia by inserting two tiny sensor pads of piezoelectric type under the screen so it could pass through somewhat when pressed. The whole thing action and audio are synchronized exclusively to simulate real button operation [22].

- A central workstation is used through which surgeons does the operations in a mixture of venues; where the local staff nurse does the tasks like setting up of the machine and patient orientation.
- In textile sector, user could learn and experience the texture and eminence of fabric during the sale of cloth all the way through internet.
- The technology can future be joint with holography, generating images, and AI that would act in response to subjects.
- Haptics could be in the smart home automation system, where the technology can manage kitchen appliances from side to side force feedback, using a blend of augmented reality and sensors. It can also be used finally to be in charge of home computerization hubs, smart thermostats, illumination and more.
- Some applications, where terrain or touch information needs to be delivered, haptics may be the suitable and major competent communication channel.
- Haptics are becoming very trendy as an essential part of virtual reality systems. Frameworks are constantly fashioned to make use of haptic interfaces for 3D to give specialists a virtual experience of genuine intelligent displaying. Instances such as simulators, control systems, devices and specialized models that permit for touch-based interface with computers.
- Disney Research developed REVEL a project texture on smooth objects is mainly dynamic. It made possible for the user to feel the material greater prior to online purchase. It works perfectly on any surface, including, tabletops, human skin, walls or furniture through vibrators implanted in a chair. It gives the player look and feel as if the user is driving a race car, inclusive with collisions and skids [23, 24].

10 Summary of Haptics, Challenges and Prospective Research Potential

Since high computational power computers, large-precision automatic sensors and actuators are available in market, it has become very easy to now apply deal with over trial variables as it was not at all done before. Research states that even auditory and visual learners are benefited a lot from actions that use sense of touch. A case study report confirms that by the usage of haptic technology in pedagogy, students from middle and high school are gaining interest and in depth standing in science and associated topics. Haptics is used to communicate with objects like nanoparticles or viruses which are not possible in traditional learning system. Furthermore, beyond students are entertained to sense 3D cell rendering, gravity and friction and many more complicated topics.

Even though companies like Novint and Nokia are manufacturing interface products for haptics, one of the major challenges for the boom of this particular technology is the cost. Presently, interaction with holograms and objects with extended distance, and how far they can control and master this sort of tactile interaction is yet another major challenge. Researchers are already exploring it. Greater the result, broader is the expansion in all other industries.

Another major is the lack of ability to simply animate the focal sensory system to substitute a touch happening. For instance, cerebrum introduces nerve stimulation, and mechanical boundary that works parallel to our bodies. The abilities of haptic devices connected with human exposure are very limited. A development program should be introduced to for haptic interfaces to take account of research in the following:

a. Human haptics
b. Technological advancement
c. Correlation between humans and haptic devices.

Joystick and gloves are sufficient enough for applications which need low resolution. Other multifaceted application which uses force and tactile displays needs some more extensive research.

In view of the fact that evolution in three areas such as devices, human performance and interaction between the both are interdependent for the growth for demanding applications.

References

1. M.A. Srinivasan, *What is Haptics? Laboratory for Human and Machine Haptics: The Touch* (Lab Massachusetts Institute of Technology, 1995)
2. Er.I. Rasheed, Haptic technology: a touch revolution. Int. J. Adv. Res. Sci. Eng. **7**, 1476–1483 (2018)
3. https://www.dialog-semiconductor.com/eccentric-rotating-mass-motor-control
4. Haptic Interfaces. National Research Council, *Virtual Reality: Scientific and Technological Challenges* (The National Academies Press, Washington, DC, 1995)
5. https://www.mepits.com/tutorial/284/biomedical/applications-of-haptic-technology
6. https://www.insidescience.org/news/haptic-technology-makes-you-feel-things-aren%E2%80%99t-there
7. M. Goyal, D. Saproo, A. Bagashra, K.R. Dev, Haptic technology based on touch. Int. J. Sci. Res. Eng. Technol. (IJSRET) **2**, 468–471 (2013)
8. S.N. Meshram, A.M. Sahu, Haptic science and technology in surgical simulation, medical training and military application. Int. J. Comput. Sci. Mob. Comput. (IJCSMC) **3**(4), 156–165 (2014)

9. L.L. Lien, Y.H. Chen, Haptic surgical simulation: an application to virtual suture. Comput. Aided Des. Appl. **3**, 203–210 (2006)
10. C. Basdogan, S. De, J. Kim, M. Muniyandi, H. Kim, M.A. Srinivasan, Haptic in minimally invasive surgical simulation and training. IEEE Comput. Graph. Appl. 56–64 (2004)
11. J. Pettersson, K.L. Palmerius, H. Knutsson, O. Wahlström, B. Tillander, M. Borga, Simulation of patient specific cervical hip fracture surgery with a volume haptic interface. IEEE Trans. Biomed. Eng. **55**(4) (2008)
12. L.-M. Liu, W. Li, J.-J. Dai, Haptic technology and its application in education and learning, in *International Conference on Ubi-Media Computing and Workshops (Ubi-Media)* (2010)
13. K. McCann, Bringing haptic technology to automobiles. AutoTechnology (2002). http://www.springer.com/978-3-642-22657-1
14. N.F. Vaibhav, V.S. Mohit, A.A. Anilesh, Application of haptic technology in advance robotics, in *INTERACT—2010*, 2010, pp. 273–277
15. Texas Instruments, http://www.ti.com/lit/ml/slyt539/slyt539.pdf
16. Precision Microdrives, https://www.precisionmicrodrives.com/vibration-motors/linear-resonant-actuators-lras
17. https://www.precisionmicrodrives.com/application-notes/ab-020-understanding-linear-resonant-actuator-characteristics
18. https://www.prnewswire.com/news-releases/haptic-technology-applications-and-global-markets-report-2017-300484007.html
19. S. Sri Gurudatta Yadav, R.V. Krishnaiah, Haptic science and technology. Int. J. Comput. Eng. Appl. **2**, 139–146 (2014)
20. S.S. Oo, N.H.H.M. Hanif, I. Elamvazuthi, Closed-loop force control for haptic simulation: sensory mode interaction, in *Proceedings of 3rd IEEE Conference on Innovative Technologies in Intelligent Systems and Industrial Applications (CITISIA 2009)*, July 2009, pp. 96–100
21. https://www.dmegc.de/files/New_DMEGC_Haptic_Device_for_Automotive.pdf
22. L. Immonen, Haptics in military applications, Seminar paper, Department of Computer Sciences. Interactive Technology, University of Tampere, Dec 2008, 14 pages
23. A. Alur, P. Shrivastav, A. Jumde, Haptic technology—a comprehensive review of its application and future prospects. Int. J. Comput. Sci. Inf. Technol. (IJCSIT) **5**(5), 6039–6044 (2014)
24. M. Sreelakshmi, T.D. Subash, Haptic technology: a comprehensive review on its application and future prospects. Mater. Today Proc. **4**(2), 4182–4187 (2017)

Detection and Behavioral Analysis of Preschoolers with Dyscalculia

Sheffali Suri, Annu Tirkey, J. Delphy and Sagaya Aurelia

Abstract Human behaviours are influenced by various factors that might impact their thought process. The way human beings response in situations have a strong connection with genetic makeup, cultural values and experiences from the past. Behaviour Analysis discusses the effect of human response to external/internal stimuli. This study helps in understanding behaviour changes among individuals suffering from various psychological disorders. Dyscalculia is one similar type of learning disorder [LD] which is commonly found among individuals and goes undetected for years. It is a lifelong condition which causes difficulty for people to perform mathematics-related tasks. Dyscalculia is quite eminent at every age. Since the symptoms are prominent from a young age, it can be detected at the earliest. Dyscalculia has no medical treatment but can be minimized by getting involved in some brain exercises especially created for children with Learning Disabilities. The chapter deals with minor research and the behaviour analysis for the above-mentioned disorder among pre-schoolers. In this chapter, a study of the behavioural patterns of pre-schoolers with dyscalculia is performed. This chapter also attempts to propose a model that can detect and predict the possibility of a child suffering from dyscalculia. It also includes a number of brain training activities that can help them to improve and enhance their confidence in mathematics.

Keywords Behaviour analysis · Behaviour · Dyscalculia · Analysis · Brain training

---

S. Suri (✉) · A. Tirkey · J. Delphy · S. Aurelia
Department of Computer Science, CHRIST (Deemed to be University), Bangalore, India
e-mail: sheffali.suri@cs.christuniversity.in

A. Tirkey
e-mail: annu.tirkey@cs.christuniversity.in

J. Delphy
e-mail: delphy.j@cs.christuniversity.in

S. Aurelia
e-mail: sagaya.aurelia@christuniversity.in

D. J. Hemanth (ed.), *Human Behaviour Analysis Using Intelligent Systems*, Learning and Analytics in Intelligent Systems 6,
https://doi.org/10.1007/978-3-030-35139-7_3

1 Introduction

The brain is the most mysterious organ in a human body. There are many theories related to the brain which are yet to be discovered by the researchers. With the advancement in technology, scientists can investigate the study of the human brain in ways that were not possible in the past. Various researches have been carried out to decode the brain and its impact on our behaviour. These researches focus on understanding how different parts of the brain commands human behaviour. With the idea of the brain controlling human behaviour pattern, important discoveries about how the brain works and how dysfunctions of different parts of the brain may result in several disorders.

Neurodevelopmental disorders are one such category of disorders caused when the development of the central nervous system is interrupted. In a layman'sterms, it is the defect that may take place at early development stage of brain which may lead to certain disabilities like attention-deficit/hyperactivity disorder (ADHD), autism, learning disabilities, intellectual disability (also known as mental retardation), conduct disorders, cerebral palsy, and impairments in vision and hearing. One of the most common Neurodevelopment disorders among children is Learning Disabilities (LD). Further researches are being carried out concerning the identification of dyscalculia among children and its impact on their behaviour patterns.

The following sections would discuss Behavioural Analysis, behavioural disorders and its types. Behaviour can be defined as how one acts or acts particularly towards others and Analysis can be defined as the way towards breaking an intricate subject or substance into littler parts to pick up a superior comprehension of it. So together Behaviour analysis is a process that includes both a basic and applied science. It's origin, as science go back, is almost a century old, and it's applied focus is more than a half-century old. Training in the field occurs at the bachelors, masters, and doctoral level. For many years, this training was mainly available only in departments of psychology. In the more recent years, applied behaviour analysis programs are also found in colleges of education and stand-alone behaviour analysis programs. Behaviour analysis is an exact method to manage to understand the lead of individuals and different creatures. It acknowledges that all practices are either reflexes conveyed by a response to explicit lifts in nature, or a consequence of an individual's past/history, including especially stronghold and order, together with the individual's present convincing state and controlling updates.

It merges parts of thinking, framework, and mental theory. It was created in the late nineteenth century as a reaction to significance cerebrum examine and other ordinary sorts of mind science, which regularly experienced issues making estimates that could be attempted likely. The most punctual subsidiaries of Behavior can be followed back to the late nineteenth century where Edward Thorndike spearheaded the law of impact, a procedure that included fortifying or debilitating conduct using fortification and discipline.

Behavioural analysis method uses machine learning, artificial intelligence, big data, and analytics to identify malicious, stealth behaviour by analysing subtle differences in normal and abnormal activities in order to proactively stop cyber attacks before the attackers have the ability to fully execute their destructive plans and has proven to be a particularly effective learning tool for helping elderly people and children with learning or developmental delays acquire and maintain new skills.

These treatments include ABA (applied behaviour analysis) and utilize techniques such as discrete trial training. The basic principles of behaviour medication are often adapted for use in educational settings, the workplace, and childcare.

All young children may be naughty, defiant and impulsive from time to time, which is perfectly normal. In many case, a few kids have very troublesome and testing practices that are outside the standard for their age [1] (Fig. 1).

The most common disruptive behaviour disorders include:

- oppositional defiant disorder (ODD)
- conduct disorder (CD)
- attention deficit hyperactivity disorder (ADHD) and
- Learning disability (LD).

These four behavioural issues share some basic side effects, so a determination can be troublesome and tedious. A kid or a youngster may have two issues in the

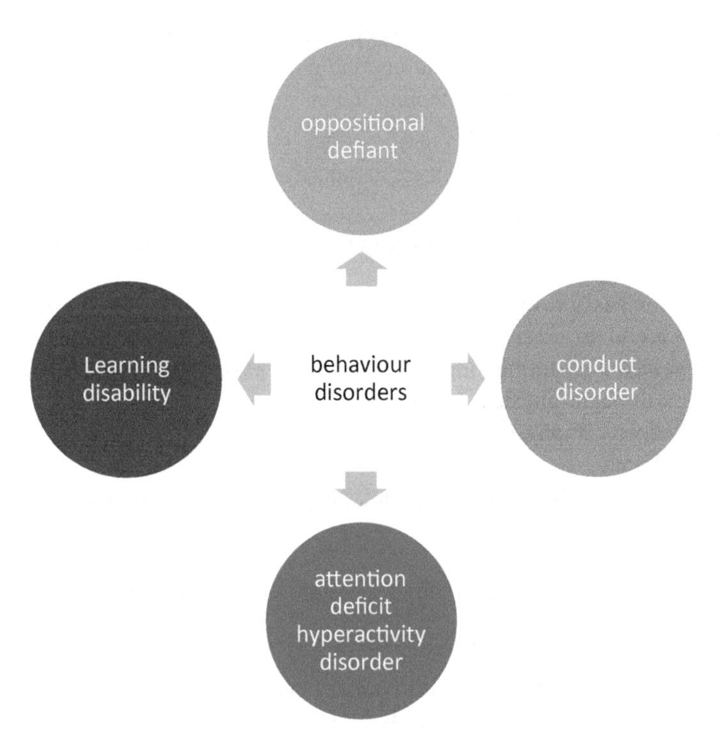

Fig. 1 Behaviour disorders

meantime. Other fuelling elements can incorporate passionate issues, state of mind issue, family challenges and substance misuse [1].

(a) **Oppositional defiant disorder**

Around one in ten children, under the age of 12 years are thought to have an oppositional defiant disorder (ODD), with boys outnumbering girls by two to one.

Some of the typical behaviours of a child with ODD include:

- Easily incensed, irritated or bothered
- Frequent fits
- Argues as often as possible with grown-ups, especially the best-known grown-ups in their lives, for example, guardians
- Refuses to obey rules
- Seems to purposely attempt to bother or disturb others
- Low confidence
- Low disappointment edge
- Seeks to censure others for any setbacks or offences [1].

(b) **Conduct disorder**

Children with conduct disorder (CD) are often judged as 'bad kids' because of their delinquent behaviour and refusal to accept rules. Around 5% of 10-year-olds are thought to have CD, with boys outnumbering girls by four to one. Around one-third of children with CD also have attention deficit hyperactivity disorder (ADHD).

Some of the typical behaviours of a child with CD may include:

- Frequent refusal to obey guardians or other specialist figures
- Repeated truancy
- Tendency to utilize drugs, including cigarettes and liquor, at an all-around early age
- Lack of compassion for other people
- Being forceful to creatures and other individuals or demonstrating perverted practices including tormenting and physical or sexual maltreatment
- Keenness to begin physical battles
- Using weapons in physical battles
- Frequent lying
- Criminal conduct, for example, taking, purposely lighting flames, breaking into houses and vandalism
- A propensity to flee from home
- Suicidal propensities—even though these are progressively uncommon [1].

(c) **Attention deficit hyperactivity disorder**

Around 2–5% of children are found to have attention deficit hyperactivity disorder (ADHD), with boys outnumbering girls by three to one.

The characteristics of ADHD can include:

- Inattention—difficulty concentrating, forgetting instructions, moving from one task to another without completing anything.
- Impulsivity—talking over the top of others, having a 'short fuse', being accident-prone.
- Over activity—constant restlessness and fidgeting [1].

(d) **Learning disability**

Learning disability is a neurological issue. In basic terms, taking in incapacity results from a distinction in the manner an individual's mind is "wired", children with learning inabilities are as savvy as or more intelligent than their companions.

Be that as it may, they may experience issues perusing, composing, spelling, thinking, reviewing and additionally sorting out data whenever left to make sense of things without anyone else's input or whenever educated in ordinary ways.

A learning inability can't be restored or fixed; it is a long-standing issue.

With the correct help and mediation, notwithstanding, youngsters with learning inabilities can prevail in school and go on to effective, frequently separated professions further down the road (Fig. 2).

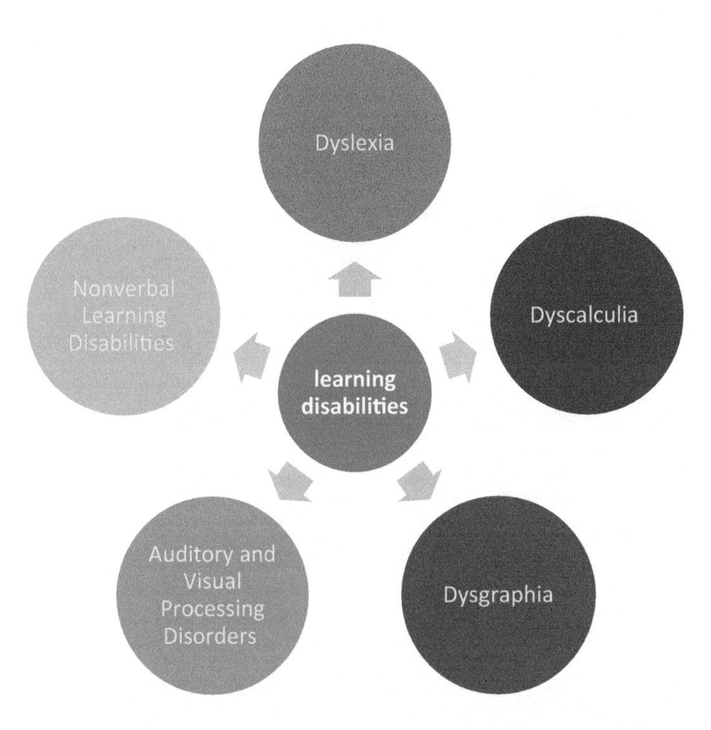

Fig. 2 Learning disabilities

Common learning disabilities

- Dyslexia—a language-based disability in which a person has trouble understanding written words. It may also be referred to as a reading disability or reading disorder.
- Dyscalculia—a mathematical disability in which a person has a difficult time solving the arithmetic problem and grasping math concepts.
- Dysgraphia—a writing disability in which an individual finds it hard to form letters or write within a defined space.
- Auditory and Visual Processing Disorders—sensory disabilities in which a person has difficulty understanding language despite normal hearing and vision.
- Nonverbal Learning Disabilities—a neurological disorder which originates in the right hemisphere of the brain, causing problems with visual-spatial, intuitive, organizational, evaluative and holistic processing functions.

The causes of ODD, CD, ADHD and LD are unknown but some of the risk factors include:

- Gender
- Gestation and birth
- Temperament
- Family life
- Learning difficulties
- Intellectual disabilities
- Brain development [1, 2] (Fig. 3).

Fig. 3 Causes of ODD, CD, ADHD and LD

2 Dyscalculia

Dyscalculia can cause extreme trouble in making arithmetical estimations, because of mental disorder. Dyscalculia is a particular learning incapacity in math. Children with dyscalculia may experience issues understanding number-related ideas or utilizing images or capacities required for achievement in mathematics.

Dyscalculia is a specific learning insufficiency in mathematics. Youngsters with dyscalculia may experience issues understanding number-related thoughts or using pictures or limits required for achievement in arithmetic [3]. Dyscalculia is a commonplace learning issue that impacts youngsters' ability to do the math. It does not impact them at school, in any case.

It causes inconveniences in regular day to day existence. The inspiring news is some various sponsorships and systems can enable youngsters to pick up the capacities they need.

The people suffering from this disorder generally have difficulty in differentiating quantitative concepts. They will not be able to find the difference between numbers and words even if they are same. For example, the patient will not be able to identify that number 10 and the word "ten" are the same. Kids with this kind of disorder will not be able to solve the mechanisms of math problems even if they recall the steps involved in it.

The major problem faced in this situation is, the kid will be unable to find or recall when and where the formulae to be implemented. They also find difficulty in memory retention of numbers for long while solving the mathematical problems called working memory.

2.1 *History of Dyscalculia*

A Swedish nervous system specialist who found that it was feasible for an individual to have debilitated scientific capacities that did not influence knowledge in general in 1919. It is additionally alluded to as formative math issue and "number visual impairment". Along these lines, the "idea of dyscalculia came to existences".

An investigation was first directed on youngsters in 1974 by Ladislav Kos in Bratislava on dyscalculia. His discoveries demonstrated that Dyscalculia is an auxiliary issue of scientific capacities. Later numerous investigations were directed on dyscalculia and different outcomes were yielded concerning the side effects of dyscalculia, the issues confronted by kids in learning math and covering of dyscalculia with other learning handicaps.

2.2 Origin of Dyscalculia

Intellectual sciences have not built up an intensive comprehension of the beginnings of dyscalculia. In any case, there are a few revelations that have built up some significant connections amongst dyscalculia and other learning and mental issue. Numerous investigations have demonstrated that an extensive number of dyscalculic youngsters endure simultaneously from extra determined troubles.

For instance, one investigation found that about 17% of dyscalculic youngsters are likewise dyslexia, and another 26% experienced the impacts of consideration shortage hyperactivity issue [4]. After two years, it was found that 4 of every 10 individuals who are determined to have dyslexia additionally experience difficulty with science to a few quantifiable degrees [4]. An autonomous determination of dyscalculia is frequently hard to discover [4].

Dyscalculia disorder is found all over the world. It is estimated that about 5% of the total population in the world are suffering from this disease. When it comes to India, about 5.5% of kids are suffering from this disorder. Dyscalculia is found more in school children than in elders from the year 2017 to 2019. Dyscalculia signs and symptoms are identified in four different stages namely in preschool, grade school, middle school and high school. But this disorder is mainly observed in preschool when compared to other stages. Kids with this kind of disorder find difficulty in counting the numbers and sorting them [5].

2.3 Symptoms Concerning Age Group

(a) The symptoms in the preschool are:

- Experiences difficulty figuring out how to check and skirt numbers long after children a similar age can recollect numbers organized appropriately.
- Battles to perceive designs, for example, littlest to biggest or tallest to the briefest.
- Experiences difficulty perceiving number images.
- Doesn't appear to comprehend the significance of checking. For instance, when requested five squares, she just gives you an armful, as opposed to forgetting about them.

(b) The symptoms that occur at grade school are:

- Encounters issues learning and inspecting major math realities, for ex-plentiful, 2 + 4 = 6.
- Battles to recognize +, − and various signs, and to use them adequately.
- May regardless use fingers to count instead of using additionally created strategies, as mental math.

- Battles to grasp words related to math, for instance, more unmistakable than and not actually.
- Experiences trouble with visual-spatial depictions of numbers, for instance, number lines.

(c) The symptoms that occur at middle school are:

- Experiences issues understanding spot regard.
- Experiences trouble forming numerals unquestionably or putting them in the correct fragment.
- Experiences trouble with segments and with assessing things, like fixings in a fundamental equation.
- Fights to monitor who's triumphant in diversions entertainments.

(d) The symptoms that occur at high school are:

- Battles to apply math ideas to cash, including evaluating the all-out cost, rolling out precise improvement and making sense of a tip.
- Experiences considerable difficulties getting a handle on data appeared on diagrams or outlines.
- Experiences issues estimating things like fixings in a straightforward formula or fluids in a container.
- Experiences difficulty discovering various ways to deal with a similar math issue (Fig. 4).

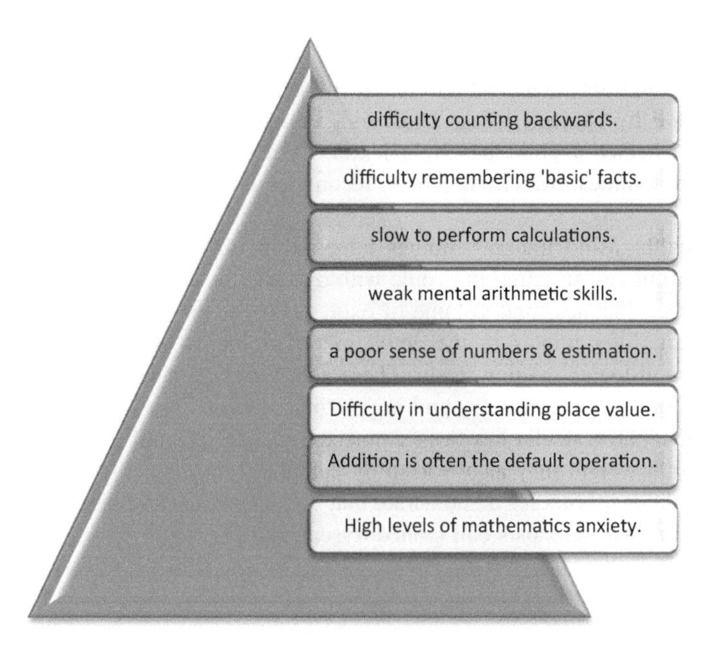

Fig. 4 Common symptoms

Typical symptoms include:

- Difficulty counting backwards.
- Difficulty remembering 'basic' facts.
- Slow to perform calculations.
- Weak mental arithmetic skills.
- A poor sense of numbers and estimation.
- Difficulty in understanding place value.
- Addition is often the default operation.
- High levels of mathematics anxiety.

There is no solution for dyscalculia. It is anything but a stage that a child can outgrow. By the time most youngsters are determined to have dyscalculia; they have an insecure math establishment. The objectives of determination and treatment are to fill in whatever number of holes as it could reasonably be expected and to create ways of dealing with stress that can be utilized all through life. There are sure tests directed to analyze if a kid is suffering from dyscalculia or not.

2.4 *Possible Causes of Dyscalculia*

The main cause for the dyscalculia disorder is not identified by any researchers till now. But some main factors affect human brain structure and functionality causing dyscalculia [3]. They are the environment, genes, mental health, and cerebrum damage and so on.

Genes: Research shows that a bit of the qualification in youngsters' math scores can be explained by characteristics. Thus, differentiates in innate characteristics may influence whether a child has dyscalculia. Dyscalculia will all in all continue running in families, which moreover recommends that characteristics accept a vocation.
Mental health: Mind imaging studies have exhibited a couple of differences in cerebrum limit and structure in people with dyscalculia. The refinements are in the surface zone, thickness and volume of explicit bits of the brain. There are in like manner that differentiates in the activation of domains of the mind related to numerical and logical getting ready. These zones are associated with key learning aptitudes, for instance, memory and organizing.
Environment: Dyscalculia has been connected to fatal alcohol disorder. Rashness and low birth weight may also lead to dyscalculia.
Cerebrum damage: Studies demonstrate that damage to specific pieces of the mind can result in what scientists call gained dyscalculia [3].

2.5 Dyscalculia in Children

Kids with dyscalculia think that it is difficult to learn maths procedures that are instructed at school, such as including addition, subtraction, multiplication and division. They also find difficulty in differentiating largest and smallest numbers.

The symptoms that are shown in children are:

Poorly coordinated correspondence: When we check out coins on a table, the greater part of us move each coin unmistakably with a finger as we separate it from the heap. An early indication of dyscalculia is that the kid doesn't contact or move the coins while attempting to count them.

Inability to perceive little amounts: Early number sense means seeing early number names and appending a mark to little amounts. For instance, if we place some 4 number of cards and ask the children to arrange them in order, it will be difficult for them.

2.6 Other Related Disorder that Occurs with Dyscalculia

Children with learning and consideration issues frequently have more than one issue. There are a couple of issues that frequently appear with dyscalculia. Moreover, the symptoms of other disorders also appear like the symptoms of dyscalculia which enables the parents to differentiate among the two disorders [6].

Testing for dyscalculia has to be done as a major aspect of a full evaluation. That way, some other learning and consideration issues can be received at the equivalent time represented as items or digits. This is detected by making the child perform addition, subtraction and other mathematical problems [6].

Some major disorder that appears with dyscalculia is:

Dyslexia: Kids most of the times, have both dyslexia and dyscalculia. Scientists have discovered that 43–65% of children with math inabilities will also have reading disabilities which is nothing but dyslexia.

ADHD: Dyscalculia and ADHD frequently happen in the meantime. In some cases, children will make math blunders given ADHD challenges. They may experience difficulty focusing on detail.

Official working issues: Executive capacities are key aptitudes that affect learning. They incorporate working memory, adaptable reasoning, and arranging and sorting out. Shortcomings in these zones can make math troublesome.

Math tension: Kids with math problems are so stressed over the possibility of doing the math that their dread and apprehension can prompt horrible showing on math tests. A few children may have both math tension and dyscalculia [6].

2.7 Identifying Dyscalculia

Dyscalculia is unmistakable and hard to recognize precisely. In any case, it is critical to perceive that similarly as understudies may have "dyslexic inclinations" so they will have maths troubles; (dyscalculia inclinations) Dyscalculia students take a shot at fundamental math which is the essential marker. It partners with trouble in numbers. They have an absence of certainty.

In their math assignments, Dyscalculia understudies may display challenges in:

- reviewing timetable
- recognizing images
- doing figuring
- discovering spot esteem
- math vocabulary
- perceiving reverse tasks
- sequencing
- accomplishing more activity.

2.8 Types of Dyscalculia

Dyscalculia is seen in kids in a wide range of sorts and subtypes. Here are a few sorts of dyscalculia.

Developmental dyscalculia is a particular learning inability influencing the ordinary procurement of math aptitudes. Hereditary, neurobiological, and epidemiologic proof shows that dyscalculia, like other learning handicaps, is a cerebrum based issue. Be that as it may, poor education and natural hardship have additionally been involved in its aetiology. Since the neural system of the two halves of the globe involves the substrate of typical number juggling abilities, dyscalculia can result from the brokenness of either side of the equator, although the left of temporal territory is of specific noteworthiness [3]. The predominance of formative dyscalculia is 5–6% in the school-matured populace and is as basic in young ladies as in young men. Dyscalculia can happen as an outcome of rashness and low birthweight and is habitually experienced in an assortment of neurologic issue, for example, consideration deficiency hyperactivity issue (ADHD), formative language issue, epilepsy, and delicate X disorder. For developmental dyscalculia has demonstrated to be a persevering learning incapacity, in any event for the present moment, in about portion of influenced preteen students. Instructive intercessions for dyscalculia go from repetition learning of number juggling certainties to creating methodologies for explaining math works out. The long haul guess of dyscalculia and the job of remediation in its result are yet to be resolved [3, 7, 8].

- **Formative dyscalculia**

Kids with formative dyscalculic discover hard to check

- perceive numerical signs
- compute
- Confuse with numbers.

- **Acalculia**

It is the later stage; it is because of wounds to mind or stroke. Understudy is unfit to convey out typical arithmetic like expansion, subtraction, division and increase. Verbal dyscalculia Kids find it hard to name verbally, name the images, signs and tallying of numbers and things.

- **Operational dyscalculia**

Operational dyscalculia is related to issues in applying rules during numerical activities. It likewise prompts disarray in recognizing numerical images.

- **Consecutive dyscalculia**

This issue alludes to handicap to check numbers as indicated by grouping to issue with figuring time, checking plan, following bearing and taking estimation[9, 10].

2.9 Dyscalculia and Language Difficulties in Arithmetics

Numerical thoughts can be summed up through making an association between language, images, pictures and genuine circumstance.

- Vocabulary—Specialized vocabulary of science challenges incorporates the accompanying.
- Technical words were not utilized by grade school students.
- Perception of youngsters that arithmetic and day by day life is unique.

(A) Dynamic and Natural Language

The understudies think that its hard to relate the scientific images, articulations mental taking care of issues, intelligent association to vide scope of genuine circumstance. This is a direct result of the unique nature of science language.

(B) Arithmetic Underachievers Learning Spatial Information

The idea of visual-spatial information was depicted utilizing number juggling learning. The underachievers of arithmetic experience issues in understanding numerical thoughts and basic spatial ideas and connections.

There are two sorts of spatial information in science educational plans;

1. Information of spatial connections in reality and
2. Information of how spatial connections are utilized to speak to genuine world thoughts. Use of spatial information to speak to science ideas in other zones [11].

Sorts of spatial challenges shown by underachievers. It incorporates

- challenges with directional ideas
- challenges with components as spatial ideas
- trouble to connect the spatial idea with suitable criteria
- trouble to get the shape, the idea of edge and pivot
- trouble in partner perceptual variables
- trouble in speaking to 3-D protests in 2-measurement
- perceptual troubles in segregating between the visual data and foundation data
- tangible engine inability
- understudies who have visual trouble learn by auditory [4].

2.10 Professionals Can Help with Dyscalculia

Various kinds of experts can help kids with dyscalculia in various ways. Some may work in a school setting, while others work secretly.

Here are a few kinds of experts who may support your kid:

- Specialized curriculum educators
- Math mentors or instructive advisors
- Child therapists
- Pediatric neuropsychologists.

There are no medications for dyscalculia. Furthermore, there aren't many particularly encouraging projects like there are for dyslexia. However, kids with dyscalculia may profit by multisensory guidance in math. This methodology utilizes the majority of a kid's faculty to enable her to learn abilities and get ideas. It additionally shows math ideas methodically, where one ability expands on the following. This can help kids with dyscalculia make more grounded associations with what they're realizing.

On the off chance that your child has an IEP or a 504 arrangement, she may get an assortment of backings at school. These might incorporate assistive innovation or facilities, for example, additional time on tests or having the option to utilize a number cruncher. Or on the other hand, she may have fewer issues to settle on her homework. These backings can help "make everything fair" so your child can get to what she's being instructed [6].

Your youngster can get support at school regardless of whether she doesn't have an IEP or a 504 arrangement. Her educator may give casual backings to help in class or with assignments and tests. Your child may likewise get instructional intercession through reaction to mediation (RTI) [6].

There are various systems educators use to help kids with dyscalculia. Here are a couple of models:

- Utilizing solid models that interface math to reality, for example, arranging catches. This can help fortify your child's number sense.
- Utilizing visual guides when taking care of issues. Your child may draw pictures or move around items, for example.
- Utilizing diagram paper to help keep numbers arranged.
- Utilizing an additional bit of paper to conceal a large portion of what's on a math sheet or test so your child can concentrate on one issue at any given moment [6, 12].

(A) Help Child with Dyscalculia

The things one can do:

- Investigate multisensory methods for empowering math you can use at home.
- Find programming, applications and Chrome mechanical assemblies to help your adolescent with math.
- Investigate free online assistive development contraptions for math on the web.
- Discover tabletop diversions your child can play to manufacture math capacities.
- Learn ways to deal with assistance create your child's certainty.
- See what your child can say to self-advocate in assessment school and focus school.
- Get tips on the most ideal approach to be an advertiser for your child at school [6, 13].

3 Related Works

In the present scenario, most of the researchers are studying Learning Disorder to help, support and make people aware that Dyscalculia exists, remains undetected and can be improved by practising certain algorithms, applications and games specially designed for children detected with Dyscalculia. In this section, various studies have been discussed based on various disorders and behaviour analysis for the same.

"Is your child not doing well in exams despite working hard? It could be a learning disorder." Published in The Indian Express-03-Jun-2019. And "Maths strugglers 'may have dyscalculia'", BBC News-10-Sep-2018. Below is the related research works on Dyscalculia Authors have suggested and created applications for the enhancement of the children and people suffering from Dyscalculia.

In "Identifying Dyscalculia Symptoms Related to Magnocellular Reasoning using Smartphones" author has discussed and learned about the appropriateness of portable programming as an instrument for helping conclusion of dyscalculia just as whether versatile programming can distinguish dyscalculia manifestations perhaps identified with unicellular thinking. This exploration was persuaded by the low predominance among portable programming applications produced for distinguishing basic reasons for building up the condition [14, 15].

"Towards a Mobile App Design Model for Dyscalculia Children" paper displays the written survey of a versatile application model that is identified with Dyscalculia kids and the proposed Calculic Model. Two existing portable application configuration models are identified with Dyscalculia youngsters named ERIA model and AERP model. Be that as it may, the current models are for experts to assess the viability of portable application for LD children. Calculic model planned to give versatile application configuration model to portable application fashioners and engineers in structuring and building up a versatile application for Dyscalculia kids [16].

"Development of Assistive Technology for Students with Dyscalculia". In this paper, the author has examined the reason for assistive innovation and how can it eradicate hindrances to learning for understudies with explicit learning inabilities. Assistive innovation, for example, Calculating Aid Tools: KitKanit additionally builds the understudies with dyscalculia odds of prevailing in school. The utilization of Calculating Aid Tools: KitKanit furnishes these understudies with comparable open doors as their typically learning companions. Moreover, most understudies were in concurrence with the upsides of the Calculating Aid Tools: KitKanit which could support fundamental count and math issues well and quick. They concurred that Calculating Aid Tools: KitKanit was straightforward, and not confusing. The program additionally has a wizard that can make it simpler for the understudies to ascertain and discover the appropriate responses. Additionally, the program itself is little and does not require a ton of assets on the PC. A few understudies likewise included that the program is facilitating a direct result of the utilization of activity and beautiful illustrations which help expanding their understanding when they need to ascertain. Moreover, we are attempting to cover the development level of parts, decimals, polynomial math, and geometry for the following rendition of Calculating Aid Tools: KitKanit. We in this manner expect that when the program is prepared to use by all evaluation levels of understudies with math inabilities in Thailand, the program will diminish their battle to learn science [17].

"A Brief Look into Dyscalculia and Supportive Tools". In this paper, author has examined about instruments which are either manual, PC based or online which can help youngsters experiencing Dyscalculia leaving a fourth choice not yet investigated: a portable device. These days it is the most available and favoured by the educators and understudies, because of the innovation progresses. There are no instances of this kind of hardware in the market devoted to dyscalculia, less in Portugal. Hence, this is the subsequent stage and this is the reason it is being built up as an android application at the University of Minho. Other methodologies and unquestionably with more adequacy are the few systems of mind preparing, dealing with cerebrum neuro-plasticity. As alluded previously, the portion of math ideas obtaining on the cerebrum can be created and instructed, so the degree of dyscalculia can be decreased. This is an excellent methodology of a strong instrument and, luckily, it has been spread between the mainland.

"Screening for Disorders of Mathematics via a Web Application". In this paper, author has examined about the web application screener that was created and conveyed by the present convention uncovered that youngsters with dyscalculia had essentially lower scores of right answers and bigger latencies in contrast with kids that were enlisted as correlation bunch in all assignments of the battery. There result confirms their examination theory that Greek understudies that were analyzed as dyscalculic by paper-and-pencil tests would present lower scores and bigger latencies in the errands of the web application screener. Their discoveries affirm past examinations showing that dyscalculia indications incorporate the powerlessness to comprehend the connection between numbers as far as their worth, for instance, a failure to perceive that 7 is bigger than 6. Some dyscalculics experience difficulty understanding that a number can be comprised of a mix of at least two separate numbers. They may see a number, for example, 7 as a solitary unit, and experience difficulty understanding that it can likewise be made up by the obligation of the numbers 3 and 4, or 5 and 2. Further dyscalculia side effects incorporate trouble in getting a handle on numerical activities (including, subtracting, increasing and isolating) and a battle to recollect how to do every task [18, 4].

"Learning Disability Diagnosis and Classification—A Soft Computing Approach". In this paper, authors have examined the discourses which they had with specialists, unique instructors, educators and guardians of kids with LD, which could comprehend and screen the way toward diagnosing LD in India. The system for diagnosing LD was very unwieldy and tedious. Without analysis, no therapeutic training is commonly given to youngsters, because of whom they think that it's hard to change with the customary instruction example and end up enduring with gloom or at times co-related disarranges like ADHD. We have exhibited a basic LVQ calculation which when contrasted with other delicate processing calculations gives a focused outcome. The strategy isn't just basic and simple to recreate in tremendous volumes however gives practically identical outcomes dependent on acknowledged benchmarks. Be that as it may, there is an extension for a further upgrade of the framework by finding a blend of calculations to develop a tastefully precise model. It is additionally observed that on expanding the quantity of information in the preparation set of the framework, the general precision demonstrates a promising development. Arrangement to recognize immaterial and unnecessary factors which may prompt a decline in conclusion procedure time and increment in precision. This can be advantageous for the unique instructors, specialists and educators by giving proposals that lead to the avoidance of repetitive tests and sparing of time required for diagnosing LD. All in all, the focal point of our examination is to distinguish the early finding of LD and to help specialized curriculum network in their mission to be with the standard [19, 20].

"Developmental Dyscalculia is Characterized by Order Processing Deficits: Evidence from Numerical and Non-Numerical Ordering Tasks". In this paper, authors contrasted the exhibition of youngsters and DD and kids without science troubles on a scope of assignments surveying requesting aptitudes, size handling/ estimation abilities, and hindrance. They isolated youngsters into two gatherings firmly coordinated on age, sexual orientation, financial status, instructive encounters, IQ and perusing capacity. Their discoveries uncovered contrasts between the gatherings both in requesting and greatness preparing capacities. In particular, both numerical and non-numerical requesting aptitudes were weakened in DD, just as execution on the dab examination and number line errands. By and by, the distinctions seemed, by all accounts, to be quantitative, as opposed to subjective, as separation impacts, just as other inside assignment controls had a similar impact on the exhibition of the two gatherings. A calculated relapse examination showed that a blend of the parental requesting poll, request working memory and the number line assignment could be utilized to accurately distinguish 80% of the members as dyscalculic or non-dyscalculic. In reality, the requesting undertakings alone recognized a somewhat bigger extent (82.5%) of members effectively. This has extraordinary criticalness for the improvement of novel symptomatic strategies for DD. Specifically, because even extremely youthful youngsters can play out some non-numerical requesting assignments, an early finding of vulnerability to maths challenges may be conceivable [21].

"Basic Numerical Capacities and Prevalence of Developmental Dyscalculia: The Havana Survey". This paper shows a connection between the fundamental ability to speak to and procedure accurate numerosities and the individual contrasts in arithmetical skill in a huge scope of the formative time. Inside and out investigation uncovered contrasts in the commitment of count and numerical size correlation on number juggling all through improvement until immaturity, however further research concentrated on longitudinal plans is important to affirm this. These discoveries bolster the attestation that shortages in essential numerical limits can serve to recognize DD, at school passage as well as at ages when increasingly refined math abilities are procured. Their results additionally bolster end that limit trial of BNB are exact and predictive apparatuses for recognizing youngsters with DD. Besides, the contrastive discovering of sexual orientation proportion information and pervasiveness for various subgroups firmly propose that DD because of shortages in essential number capacities could be a particular issue that includes a small amount of the low arithmetical achievers. At long last, this investigation underscores the significance of preparing procedures on low-level numerical handling and the obtaining of emblematic portrayals of numerical extents. This is noteworthy not just for the administration of atypical advancement of math abilities yet in addition to improving math learning potential in regularly creating youngsters [22, 23].

"Diagnosing Developmental Dyscalculia Based on Reliable Single Case FMRI Methods: Promises and Limitations". In this paper, the author contemplated and distinguished veering off neural instruments in the dyscalculic gathering even though no distinctions in execution were watched. The single-case examination uncovered that dyscalculic kids demonstrate a move of initiation from essential to higher visual frameworks. Extra investigations recommend that this move obliges higher actuation in the front parietal cortex which could speak to the remuneration of shortfalls in the polygonal parts of number portrayal through finger related portrayal in the DD gathering. The author contended that these distinctions in cerebrum enactment without conduct contrasts can be translated as steady compensatory neural systems that have advanced after some time. Thus the multivariate example methodologies had the option to separate the two gatherings, regardless of whether there were no distinctions in execution. The diverse cerebrum territories identified through multivariate example examination recommend that future network investigation approaches. "Dyscalculia: From Brain to Education". In this paper, the author has mentioned about the exploration in psychological and formative neuroscience which is giving another way to deal with the comprehension of dyscalculia that emphasizes a centre shortfall in getting sets and their numerosities, which is fundamental to all parts of grade-school arithmetic. The neural bases of numerosity handling have been researched in auxiliary and utilitarian neuro-imaging investigations of grown-ups and kids, and neural markers of its impairment in dyscalculia have been distinguished. New intercessions to fortify numerosity handling, including versatile programming, guarantee powerful evidence-based training for dyscalculic students.

4 Methodology

4.1 Screening Test

See Fig. 5.

- Non-verbal Test

This test would include questions regarding the non-verbal section of mathematics that comprises an identification of numbers, approximation and comparison problems. This test would evaluate numbers sense and fact retrieval i.e. the ability of the child to identify, memorise and compare values.

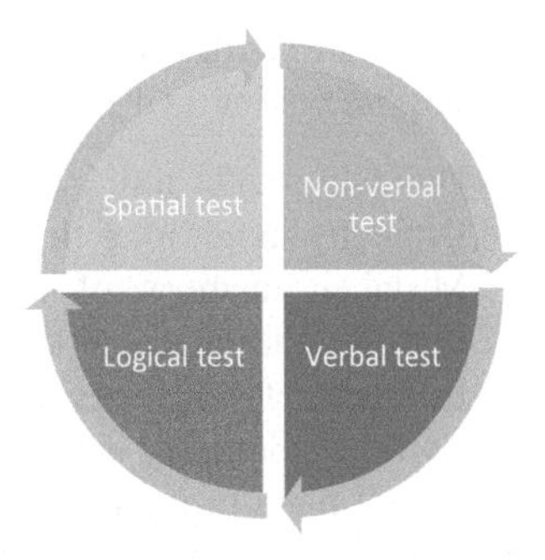

Fig. 5 Screening test

- Verbal Test

This test would include questions related to several facts. Number facts screening comprises of basic addition, subtraction and multiplication word problems which would evaluate students' ability to compute.

- Logical Test

This test is considered to be a bit difficult when compared to former ones. It includes certain higher maths as well as problem-solving questions. This test would evaluate students' ability to use strategy and procedures to solve a problem.

- Spatial Test

This test consists of problems based on the number line and basic geometry. This test would check if the student is facing difficulty with subsidizing and apprehension with spatial data.

Based on the results of these tests, It can be predicted if the child is suffering from Dyscalculia or not. Also, behaviour patterns of students (preschoolers) suffering from Dyscalculia has been analysed to understand changes.

4.2 Data

Preschool children who showed less ability to estimate the number of objects in a group was 2.4 times more likely to have a later mathematical learning disability than other young people. Our study limits to age group 3–5 as it the creating phase of Dyscalculia and prompt consideration can help to avoid the severity of a condition in future. Thus, this study would concentrate on understudies in preschool [24, 25].

5 Model and Implementation

Based on the results of the tests conducted above, the following checklist is used for scoring, screening and to analyse the behaviour patterns among students.

- Difficulty with time, headings, reviewing plans, successions of occasions. Difficulty monitoring time. Much of the time lately.
- Mistaken memory of names. Poor name-face affiliation. Substitute names starting with the same letter.
- Inconsistent outcomes furthermore, subtraction, augmentation and division. Awful at monetary arranging and cash the executives. Excessively moderate at mental math to figure sums, change due, tip, charge.
- When composing, perusing and reviewing numbers, these mix-ups may happen number augmentations, substitutions, transpositions, exclusions, and inversions.

- Inability to get a handle on and recall math ideas, rules, recipes, grouping (order of tasks), and fundamental math certainties (+−×/).
- Poor memory (maintenance and recovery) of math ideas might probably perform math activities one day, however, experience a mental blackout by the following! May most likely book work however then bombs tests.
- Unable to envision or "picture" mechanical procedures. Poor capacity to "envision or picture" the area of the numbers on the essence of a clock, the land locations of states, nations, seas, lanes, and so on.
- Poor memory for the "design" of things. Gets lost or perplexed effectively. May have a poor ability to know east from west, may lose things frequently, and appear to be inattentive.
- Difficulty getting a handle on ideas of formal music training. Trouble locate perusing music, picking up fingering to play an instrument.
- Difficulty with engine sequencing, observable in athletic execution, trouble staying aware of quickly changing physical headings like in oxygen-consuming, move, and exercise classes. Trouble with move step arrangements, muscle memory, sports moves.
- Difficulty recalling how to keep track of who's winning in recreations, such as bowling, cards, and so on. Regularly forgets about whose turn it is. Restricted vital arranging capacity for recreations like chess.
- Experiences tension during math assignments.
- Uses fingers to check. Loses track when checking. Can't do mental math. Includes with dabs or count marks.
- Numbers and math appear to be an unknown dialect [24].

Based on the above scoring checklist, it is classified if the student is suffering from Dyscalculia based on available scales Ravens CPM, WISC-III, or is it an irrational fear of math or numbers. Minor behaviour analysis was done to understand the behaviour differences observed among students suffering from dyscalculia.

6 Result and Discussion

The data and results were analysed using ANCOVA. From the above tests, it was quite evident that about 4–8% of pre-schoolers are suffering from Dyscalculia. It is quite evident that everyone in twenty people suffer from it. The research proposes that half of the youngsters with dyscalculia will likewise have dyslexia, and 20% ADHD. Thus, dyscalculia can best be characterized as a deficiency in the portrayal or preparing of explicitly numerical data.

From the above results, some of the most frequent behaviour patterns observed are:

(1) Anxiety—Math anxiety is an emotional issue involving self-doubt and fear of failing which is observed In students suffering from dyscalculia. The child ends up scoring fewer marks in-class assessments and thus starts avoiding studies leading to chronic anxiety.
(2) Social Awkwardness—It has been observed that students with dyscalculia suffer from social awkwardness and avoid making social interactions.
(3) Stress Level—Students are considered to have higher stress levels as compared to normal students as they do not get the right guidance for the same.
(4) Lack of Participation—It has been noticed that the students suffering from dyscalculia tends to develop complexes and thus avoid participating in class. This can be linked to lower self-esteem and later total lack of interest in other studies as well.

These are few patterns observed in the study conducted there are many more issues observed in the behaviour pattern of a student diagnosed with Dyscalculia.

7 Conclusion and Future Scope

The purpose of this chapter was to familiarise everyone with Human behaviours and its connection with the human brain. Various disorders were discussed in detail concerning our brain along with their symptoms. A detailed review was presented on Dyscalculia along with minor behaviour analysis.

For future work, the detailed study can be carried out which can be used to analyse data to understand the root cause of such Neurodevelopment disorders and eliminate them. There is still room for a lot of research to prevent these conditions or rather is prepared to combat them.

References

1. Anxiety in Children, www.anxietyinchildren.com
2. B.N. Verdine, C.M. Irwin, R.M. Golinkoff, K. Hirshpasek, NIH public access. J. Exp. Child Psychol. **126**, 37–51 (2014)
3. B. Butterworth, S. Varma, D. Laurillard, Dyscalculia: from brain to education. Science **332** (2011)
4. 'Shalev et al. (1997)', 'Butterworth (1999)', 'Gross-Tsur and Manor (1996)', 'Ostad (1998)' and 'Dickson et al. (1984)'
5. I. Rapin, Dyscalculia and the calculating brain. Pediatr. Neurol. **61**, 11–20 (2016)
6. P.J. Dinkel, K. Willmes, H. Krinzinger, K. Konrad, J.W. Koten, Jr., Diagnosing developmental dyscalculia on the basis of reliable single case FMRI methods: promises and limitations. PLOS ONE **8**(12), e83722 (2013)
7. O. Simsek, Use of a game-based app as a learning tool for students with mathematics learning disabilities to increase fraction knowledge/skill, June 2016
8. Understood for learning and attention issues, https://www.understood.org/en
9. A. Plerou, Dealing with dyscalculia over time, in *ICICTE*, no. 2008, pp. 1–12 (2014)

10. T. Käser et al., Modelling and optimizing mathematics learning in children. Int. J. Artif. Intell. Educ. **23**(1–4), 115–135 (2013)
11. G. Karagiannakis, A. Baccaglini-Frank, Y. Papadatos, Mathematical learning difficulties subtypes classification. Front. Hum. Neurosci. **8**, 57 (2014)
12. J. Borg, A. Lantz, J. Gulliksen, Accessibility to electronic communication for people with cognitive disabilities: a systematic search and review of empirical evidence. Univers. Access Inf. Soc. **14**(4), 547–562 (2014)
13. N. Sachdeva, A.M. Tuikka, K.K. Kimppa, R. Suomi, Digital disability divide in information society: literature review. J. Inf. Commun. Ethics Soc. **13**(3), 283–298 (2015)
14. F. Ferraz, H. Vicente, A. Costa, J. Neves, Analysis of dyscalculia evidences through artificial intelligence systems. J. Softw. Netw. 53–78 (2016)
15. R.K. Vukovic, N.K. Lesaux, The relationship between linguistic skills and arithmetic knowledge. Learn. Individ. Differ. **23**(1), 87–91 (2013)
16. M.M. Ariffin, F.A. Abd Halim, N. Abd Aziz, Mobile application for dyscalculia children in Malaysia, in *Proceedings of the 6th International Conference on Computing and Informatics*, Paper No. 099, 27 Apr 2017
17. S. Pieters, H. Roeyers, Y. Rosseel, H. Van Waelvelde, A. Desoete, Identifying subtypes among children with developmental coordination disorder and mathematical learning disabilities, using model-based clustering. J. Learn. Disabil. (2013). https://doi.org/10.1177/0022219413491288
18. J. Ismaili, E.H.O. Ibrahimi, Mobile learning as alternative to assistive technology devices for special needs students. Educ. Inf. Technol. (2016)
19. K.L. Luxy, Learning difficulties and attention deficit hyperactivity disorder. J. Psychiatry **20** (2), 1000404 (2017)
20. T. Nagavalli, P. Juliet, *Technology for Dyscalculic Children* (Salem, 2015)
21. I.C. Mammarella, S. Caviola, C. Cornoldi, D. Lucangeli, Mental additions and verbal-domain interference in children with developmental dyscalculia. Res. Dev. Disabil. (2013)
22. F.A. Aziz, H. Husni, Z. Jamaludin, Translating interaction design guidelines for dyslexic children's reading application, in *Proceedings of the World Congress on Engineering*, vol. II (2013)
23. A.M. Hakkarainen, L.K. Holopainen, H.K. Savolainen, The impact of learning difficulties and socioemotional and behavioural problems on transition to postsecondary education or work life in Finland: a five-year follow-up study. Eur. J. Spec. Needs Educ. **31**(2), 171–186 (2016)
24. Cognitive Learning-Breaking Barriers to Learning, http://www.cognitivelearning.co.za/
25. S.Z. Ahmad, A. Abdul Mutalib, Exploring computer assisted learning for low achieving children: a comparative analysis study. J. Teknol. **77**(29), 1–7 (2015)
26. A.C.G.C. Duijzer, S. Shayan, A. Bakker, M.F. Van der Schaaf, D. Abrahamson, Touchscreen tablets: coordinating action and perception for mathematical cognition. Front. Psychol. **8**, 1–19 (2017)

FPGA Based Human Fatigue and Drowsiness Detection System Using Deep Neural Network for Vehicle Drivers in Road Accident Avoidance System

D. Selvathi

Abstract Automobile Industry shares numerous accidents in our daily routine. Increasing rate of road accidents are due to driver distraction such as fatigue and lack of sleep. This work is intended solely for the implementation of fatigue and drowsiness detection system using the deep neural network in FPGA. In the proposed system, the image is preprocessed using median filtering and Viola Jones face detection algorithm for extracting the faces. Further, the features are extracted by using Local Binary Pattern analysis and the Max pooling is used to reduce the complexity level. These deep learning steps are followed by performing SVM classifier to define the status of the subject as drowsy or not. The system uses a camera to capture the real time image frames in addition with offline images of the system. The developed Vision-based driver fatigue and drowsiness detection system is a convenient technique for real time monitoring of driver's vigilance.

Keywords Deep neural network · FPGA · Drowsiness · Local binary pattern

1 Introduction

Driver fatigue detection is a car safety technology which prevents accidents when the driver is getting drowsy. Various studies have suggested that around 20% of all road accidents are fatigue-related, up to 50% on certain roads. The development of technologies for detecting or preventing drowsiness at the wheel is a major challenge in the field of accident avoidance systems. Driver inattention might be the result of a lack of alertness when driving due to driver drowsiness and distraction. Driver distraction occurs when an object or event draws a person's attention away from the driving task. Unlike driver distraction, driver drowsiness involves no

D. Selvathi (✉)
Department of Electronics and Communication Engineering,
Mepco Schlenk Engineering College, Sivakasi 626005, Tamilnadu, India
e-mail: dselvathi@mepcoeng.ac.in

D. J. Hemanth (ed.), *Human Behaviour Analysis Using Intelligent Systems*,
Learning and Analytics in Intelligent Systems 6,
https://doi.org/10.1007/978-3-030-35139-7_4

triggering event but it is characterized by a progressive withdrawal of attention from the road and traffic demands. Both driver drowsiness and distraction might have the same effects i.e., decreased driving performance, longer reaction time, and an increased risk of crash involvement [1].

The US National Highway Traffic Safety Administration has estimated approximately 100,000 crashes each year caused mainly due to driver fatigue or lack of sleep [2]. Autonomous systems designed to analyze driver exhaustion and detect driver drowsiness can be an integral part of the future intelligent vehicle so as to prevent accidents caused by sleep. Developing intelligent systems to prevent car accidents can be very effective in minimizing accident death toll. One of the factors which play an important role in accidents is the human errors including driving fatigue relying on new smart techniques; this application detects the signs of fatigue and sleepiness in the face of the person at the time of driving.

A variety of techniques have been employed for vehicle driver fatigue and exhaustion detection. Driver operation and vehicle behavior can be implemented by monitoring the steering wheel movement, accelerator or brake patterns, vehicle speed, lateral acceleration and lateral displacement. These are non-intrusive ways of driver drowsiness detection, but are limited to the type of vehicle and driver conditions [3]. Another set of techniques focuses on monitoring of physiological characteristics of the driver such as heart rate, pulse rate, and electroencephalography (EEG) [4].

Research in these lines have suggested that as the alertness level decreases EEG power of the alpha and theta bands increase, hence providing indicators of drowsiness. Although the use of these physiological signals yields better detection accuracy, these are not accepted widely because of less practicality. Physiological feature based approaches are intrusive because the measuring equipment must be attached to the driver. A third set of techniques is based on computer vision systems which can recognize the facial appearance changes occurring during drowsiness [5]. Thus, visual feature based approaches have recently become preferred because of their non-intrusive nature.

1.1 Causes of Drowsiness

Although alcohol and some medications can independently induce sleepiness, the primary causes of sleepiness and drowsy driving in people without sleep disorders are sleep restriction, sleep fragmentation and circadian factors.

Sleep Restriction or Loss. Short duration of sleep appears to have the greatest negative effects on alertness [6]. Although the need for sleep varies among individuals, sleeping 8 h per 24-h period is common, and 7–9 h is needed to optimize performance. Experimental evidence shows that sleeping less than 4 consolidated hours per night impairs performance on vigilance tasks. Acute sleep loss, even the loss of one night of sleep, results in extreme sleepiness.

Sleep Fragmentation. Sleep is an active process, and adequate time in bed does not mean that adequate sleep has been obtained. Sleep disruption and fragmentation cause inadequate sleep and can negatively affect functioning. Similar to sleep restriction, sleep fragmentation can have internal and external causes. The primary internal cause is illness, including untreated sleep disorders. Externally, disturbances such as noise, children, activity and lights, a restless spouse, or job-related duties (e.g., workers who are on call) can interrupt and reduce the quality and quantity of sleep. Studies of commercial vehicle drivers present similar findings. For example, the National Transportation Safety Board (NTSB) concluded that the critical factors in predicting crashes related to sleepiness were: the duration of the most recent sleep period, the amount of sleep in the previous 24 h, and fragmented sleep patterns.

Circadian Factors. The circadian pacemaker regularly produces feelings of sleepiness during the afternoon and evening, even among people who are not sleep deprived. Shift work also can disturb sleep by interfering with circadian sleep patterns [7].

1.2 Behavioral Methods

The methods mentioned thus far were deemed as either unreliable or very intrusive for real world applications, thus leading towards exploiting a different type of methodology based upon non-invasive observation of a driver's external state.

Head or Eye Position. When a driver is drowsy, some of the muscles in the body begin to relax, leading to nodding. This nodding behavior is what researchers are trying to detect. Research exploiting this feature has started [8]. Detecting head or eye position is a complex computer vision problem which might require stereoscopic vision or 3D vision cameras.

Yawning. Frequent yawning is a behavioral feature that tells that the body is fatigued or falling into a more relaxed state, leading towards sleepiness. Detecting yawning can serve as a preemptive measure to alert the driver. It should be noted that yawning does not always occur before the driver goes into a drowsy state. Therefore it cannot be used as a standalone feature and it needs to be backed up with additional indicators of sleepiness.

Detecting the state of the eyes has been the main focus of research for determining if a driver is drowsy or not. In particular, the frequency of blinking has been observed. The term PERCLOS (PERcentage of eyelid CLOSure over the pupil over time) has been devised to provide a meaningful way to correlate drowsiness with frequency of blinking. This measurement has been found to be a reliable measure to predict drowsiness.

1.3 Scope and Objective

The objective of this work is to classify the status of the driving person who is having drowsiness/fatigue and to develop a methodology for classification. The work also involves in implementing the method using FPGA in the real time environment, to prevent accidents due to sleeping nature. This also involves in focusing the potential area of investigation of handcrafted features (focusing on local patterns) to deep learning feature extraction and classification.

2 Literature Survey

The methods for assessing driver drowsiness are generally related to the measurement of the driver's state, driver performance and a combination of the driver's state and performance. For each method, different criteria must be analyzed, for example yawning, head position and eye closure can be studied in a driver's state measurement, while lane tracking and tracking distances between vehicles are involved in the studies of a driver's performance. Based on the result of different researches, the most accurate technique towards driver fatigue detection is dependent on physiological phenomena like brain waves, heart rate, pulse rate and respiration. But these techniques are intrusive and require the attachment of some electrodes on the driver, which are likely to cause annoyance. Therefore, to monitor the driver's vigilance, different techniques of computer vision can be used, which are natural and non-intrusive. These techniques focus on changes in a human's facial features like eyes, head and face [9].

Smith et al. proposed a fully automatic system, capable of multiple feature detections (among which is yawning), combating partial occlusions (like eye or mouth). To strengthen the attention level gained from reconstructing the gaze direction of a driver, additional aspects are tracked, e.g. yawning, which is determined by monitoring lip corners direction change as well as lip corner color change [10]

Liu et al. decided to explore the advantages of infrared camera in order to monitor driver's eyes. Some of the infrared sensors' main advantages are the ability to operate during nighttime and their lower sensitivity to light changes. The Viola-Jones algorithm is used once again to initially detect the driver's face and eyes. If the eyes are successfully located, the eye regions are extracted for further processing. The movement of the eyelids can be detected and used as a measure of eye closeness. The frequency of blinking is used as a measurement of how fatigued is the driver. The algorithm can track eyes in higher speeds after initial eye detection and is gender independent as well as resilient towards low levels of lighting [11].

The work by Wu et al. also determines driver's drowsiness by monitoring state of the eyes of a driver. Their detection process is divided into three stages: (i) initial face detection based on classification of Haar-like features using the AdaBoost method, (ii) eye detection by SVM classification of eye candidates acquired by applying radial-symmetry transformation to the detected face from the previous stage and (iii) extraction of local binary pattern (LBP) feature out of left eye candidate, which can be used to train the SVM classifier to distinguish between open eye state and closed eye state. The LBP of an eye can be thought of as a simplified texture representation of it. Both closed- and open-eye LBPs are distinguishable by two-state classifiers such as the SVM [12].

Tian and Qin proposed that combining several basic image processing algorithms can increase the performance of eye state detection system and bring it closer to real time performance. Moreover, two different methodologies are proposed for daytime and nighttime detection. In order to detect the driver's face during daytime, a combination of skin color matching with vertical projection is performed. This color matching will not work during night, therefore vertical projection is performed on the cropped face region acquired after adjusting image intensity to have uniform background. For locating the eyes, horizontal projection is applied on the previously detected face. The eye region is then converted to a binary image in order to emphasize the edges of the eyes. The converted image is processed through a complexity function which provides an estimate of its complexity level: an image containing open eyes has more complex contours than another image containing closed eyes. The difference in complexity level can be used to distinguish between those two states [13].

Lien and Lin follow the most commonly used approach of face and eye detection, feature extraction and eye state analysis based on different feature set for different eye state. In this paper they propose the computation of least correlated local binary patterns (LBP), which are used to create highly discriminate image features that can be used for robust eye state recognition. An additional novelty of the proposed method is the use of independent component analysis (ICA) in order to derive statistically independent and low dimensional feature vectors. The feature vectors are used as classification data for an SVM classifier, which provides information about the current eye state. This information is used to determine the frequency of blinking, which is then used to determine the state of the driver [14].

Lenskiy and Lee proposed a sophisticated system that monitors and measures the frequency of blinking as well as the duration of the eye being closed. In order to detect the eyes, a skin color segmentation algorithm with facial feature segmentation algorithm is used. For skin color segmentation, a neural network is trained by using RGB skin color histogram. Different facial regions, such as eyes, cheeks, nose, mouth etc. can be considered as different texture regions which can be used to subsegment given skin model regions of interest. Each segment is filtered additionally to create SURF features that are used to estimate each class's probability density function (PDF). The segmented eye region is filtered with Circular Hough transform in order to locate iris candidates, i.e., the location of the driver's eyes. To track the eyes over an extended period of time, Kalman filtering is used [15].

Tuncer et al. proposed an assistant system to track a lane, which was activated for those drivers who are not able to perform a good job of lane keeping. For lane detection and lane tracking, a camera based image processing algorithm is required, which use offline and real time hardware-in-the-loop simulations. In order to process the video frames coming from an in-vehicle camera pointed towards the road ahead, a PC is used that detects and computes the tracking of the lane. It calculated the required steering actions and send them to the Carmaker vehicle model [16].

In the drowsiness detection method proposed by Pilutti et al., driver assessment is determined in the context of a road departure warning and intervention system. In this method, the vehicle lateral position is used as the input and steering wheel position as the output in order to develop a system that is updated during driving. The driver's performance is determined by analyzing the changes in the bandwidth and/or parameters of such a model [17].

Any physical changes that occur within the body during the onset of fatigue are considered as physiological measures, which are a direct measure of fatigue. The activity in the brain can be determined by electroencephalographic measurements. The brain's level of alertness is changed by electrical activity, which allows the detection of sleepiness and different stages of sleep. In the paper proposed by Picot, a fatigue detection system is based on brain and visual activity. A single electroencephalographic channel is used in order to monitor the brain activity for drowsiness detection. The measurement of electrical activity of the brain electroencephalographic is determined by placing the electrodes on the scalp. Electroencephalographic data can be analyzed where rhythmic activities are calculated in several physiologically significant frequency bands in the frequency domain. In order to detect drowsiness, any change in α, θ and β analyzed in electroencephalographic data, for example, an increase of the α and θ activities and a decrease of the β activity. α and θ are linked to relaxed and eyes closed states and β is linked to active concentration. The main goal of the electroencephalographic based detection system is to detect important variations of activity in the appropriate frequency ranges. Visual activity is detected through blinking detection and characterization. Merging detection systems allows the fatigue detection system to detect three levels of drowsiness: "awake," "drowsy" and "very drowsy" [18].

Hayashi et al. proposed another method of driver drowsiness detection by pulse wave analysis with a neural network. Since the biological signal such as pulse wave sharply reflects a human's mental condition, this method is used to measure the driver's awareness condition. In order to reduce noise signals by the driver's movement, the sensor measuring the pulse wave was attached on a driver's wrist. Three indexes from the obtained pulse wave is evaluated such as sympathetic nerve activity and parasympathetic nerve activity from Heart Rate Variability analysis and is given as an input to the neural network to determine the driver's state of drowsiness [19].

Behaviors indicates the tiredness or other unsafe driving situations such as distraction take the form of yawning, blinking, eyes closure, head movements, use of mobile devices and eye glance patterns. The first step towards drowsiness detection based on behavioral features is to detect the driver's face. In this case, the search area for any facial feature is reduced to the face region. There are numerous techniques towards face detection processing images containing faces have been developed in different research categories such as face recognition, face tracking, pose estimation and expression recognition. The objective of face detection is to recognize all image regions that contain a face without considering its position, orientation and lighting conditions.

The goal of the feature invariant approaches is to find structural face features such as eyebrows, eyes, nose, mouth and hairline, which persist under various poses, viewpoints or lighting and use those features to detect faces. Such features are mostly extracted using edge detectors. A method to identify the face from a cluttered background based on segmentation is proposed. The canny edge detector and heuristics are used as an edge map to remove and group edges. Then, the ellipse is fitted to the boundary between the head region and the background and the face is detected [20].

Han et al. proposed a method of face detection based on a morphology technique to perform eye segmentation since eyes and eyebrows are the salient and relatively stable features in the human face. The located eye analogue is used to search for the potential face regions with a geometrical combination of eye, nose and mouth. A trained back propagation neural network get all potential face images and verify the face location [21].

In the method proposed by Ying et al. [22], skin color is considered as the most important feature that can be separated from other parts of the background by using the maximal varieties variance threshold. In their method, an initial patch of skin color is used to initiate the iterative algorithm. In order to detect skin color, the method presents a control histogram, which is applied on different patches in the image and the current histogram for comparison. Then, the threshold value is assigned to compare with the result from histogram comparison to analyze the skin region.

Tsai et al. proposed a method of face detection using eigenface and neural networks. In this method, the eigenface algorithm is utilized to find the face regions candidates. The neural network with the back propagation (BP) algorithm is then used to examine the face candidates. In this system, the input vector of the neural network is the projection weights of the face space. At the end, template based face verification is used to confirm the face region from the output of the neural network [23].

Different methods of driver's fatigue detection are implemented by other researchers specifically focused on changes and movement in the eyes. These techniques analyze changes in the driver's direction of gaze, eye closure and blinking rate. Another drowsiness detection method based on eyelid movement was proposed by Liu et al. In their method based on the eyelid changes from a temporal differences image, the fatigue situation is analyzed. The number of white pixels can be used for the fatigue judgment criterion in the first step. Then, the number of pixels with positive change in the three level difference image and the number of pixels with negative change between current frame and previous frame is used to represent eyelid movement from open to closed, which is useful as an indicator of drowsiness [24].

Omidyeganeh et al. used a method of fatigue detection by applying the Structural Similarity Measure to find the eye location. In their method, structural similarity measure value is evaluated between −1 and 1. When two images are the same, the max gained value is 1 and when there are some differences, the result is −1. Then the horizontal and vertical projection is applied on the eye region to determine the amount of eye closure and align the detected eye region [25].

Jimenez et al. described a new method of fatigue detection in drivers based on the percentage of closing eyes and detection of yawning and nodding. After finding the face region by using the Viola-Jones theory, the eye and mouth is located based on the candidate region of interest in the face area. After converting the image into gray scale, the threshold that was assigned by using the histogram of each eye location is applied on the area. The eye state is determined by analyzing the histogram of white sclerotic and identifying the shape of eye upper curvature compared to the eyebrow. The yawning condition is detected first by threshold calculation after histogram analysis by seeking the maximum concentration of pixels. When the amplitude ratio of the mouth doubles its minimal size, the yawning situation is found [26].

Kwok Tai Chui et al. proposed drowsiness detection process using ECG Signals of driver [27]. Smart watch based Wearable EEG System for Driver Drowsiness Detection was developed by Li et al. [28]. Rohit et al. also considered EEG signal for drowsiness detection [29]. Wearable glass with IR sensor was developed for drowsiness or fatigue detection [30]. George et al. compared the classification results of mammographic abnormality detection for the variants of local binary patterns. They suggested that although several studies have focused on deep learning features for mammographic density classification, the use of handcrafted features (focusing on intensity and local patterns) to deep learning feature extraction and classification is a potential area of investigation [31].

In this work, fatigue and drowsiness detection using handcrafted local pattern features to deep neural network with SVM Classifier based facial image analysis and implement in the FPGA hardware are focused to classify the status of the driving person who is having drowsiness/fatigue. The work involves in implementing the method using FPGA in the real time environment to prevent accidents due to fatigue and sleeping nature.

3 Proposed System

The proposed methodology is to classify the status of the driving person as drowsy or non drowsy. The overall block diagram for method is shown in Fig. 1 which consists of different modules of image database, Preprocessing work, LBP based feature extraction, Max pooling based Feature extraction and the SVM based classification.

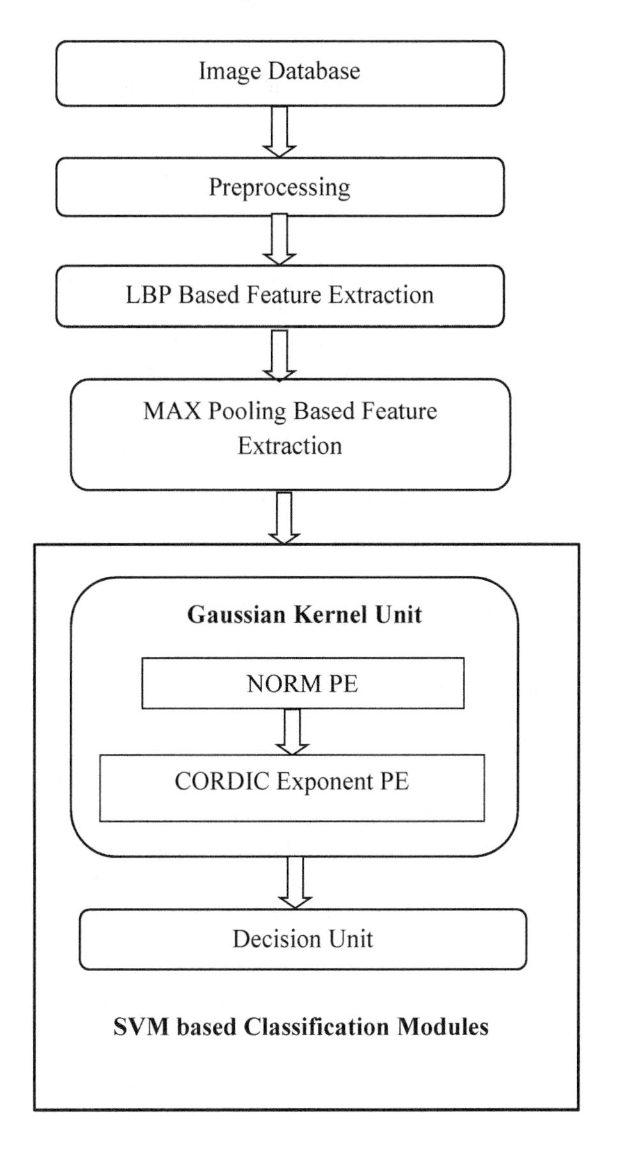

Fig. 1 Block diagram of the proposed methodology

3.1 Image Database

Due to the lack of easy availability of the standard datasets for the driver fatigue detection, the dataset obtained from the Google images has been considered as the dataset for the work and also the real time images are captured by interfacing the camera with the FPGA and those images are also used as dataset. The dataset includes the various skin tone, eye size, fatigue level and different facial structure. Totally 100 images in which 50 drowsy images and 50 non drowsy images. Among these, 50% of the dataset are used to train the classifier and the remaining 50% is used to evaluate the performance of the trained classifier.

3.2 Pre-processing

The input image taken from the database is subjected to preprocessing to improve the results of later processing. The preprocessing involves the filtering process which is done in order to reduce the noise level in the input images. In this work, the median filter is used which is a nonlinear digital filtering technique and it preserves edges while removing noise. For reducing the complexity level in the feature extraction process, the face region of the subject has been extracted using the Viola Jones face detection algorithm and the image pixel values are stored as 1-D vector for the further processing.

3.3 LBP Based Feature Extraction Process

Local binary pattern (LBP) is a type of feature extraction process, used for classification. The local binary pattern generates the pattern for identifying the pixel information of the image. The LBP operator assigns a value to every pixel of an image in a 3 × 3 neighborhood of a center pixel by comparing it with the center pixel value. If the center pixel is greater than the neighborhood pixel, it assigns "1", otherwise "0". The combined result of binary number over the window is converted to decimal number and replaces the center pixel. Figure 2 represents the extraction of LBP operator. The value of the LBP can be estimated by using Eqs. (1) and (2)

$$LBP = \sum_{p=0}^{p-1} s(x)2^p \tag{1}$$

$$S(x) = \begin{cases} 1 & \text{x > threshold} \\ 0 & \text{otherwise} \end{cases} \tag{2}$$

where s(x) is the binary value obtained by comparing the neighborhood pixel with the center pixel over the region p = 0–7 for 3 * 3 region. Here, the sliding window concept is involved to perform the computation of an LBP function which is shown in Fig. 3.

The frame size of 3 * 3 is shifted with the overlapping manner and it is preceded for the entire image computation. This results in the local texture and global shape per image is obtained. The functional diagram for an LBP based feature extraction is given in Fig. 4.

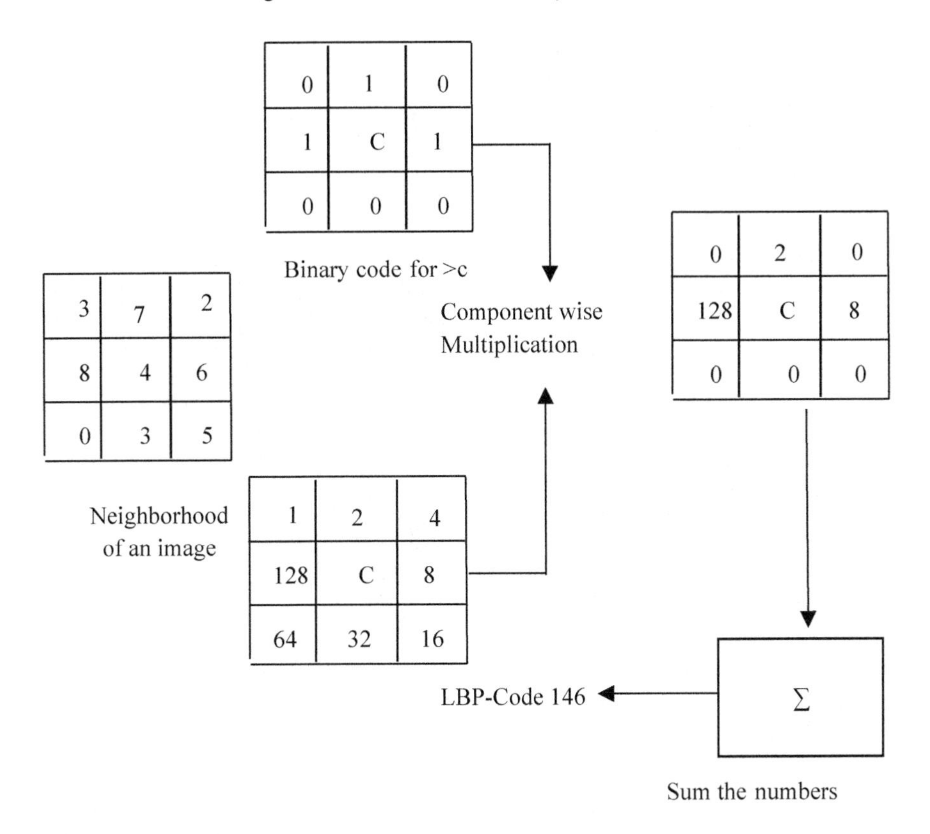

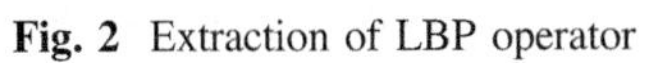

Fig. 2 Extraction of LBP operator

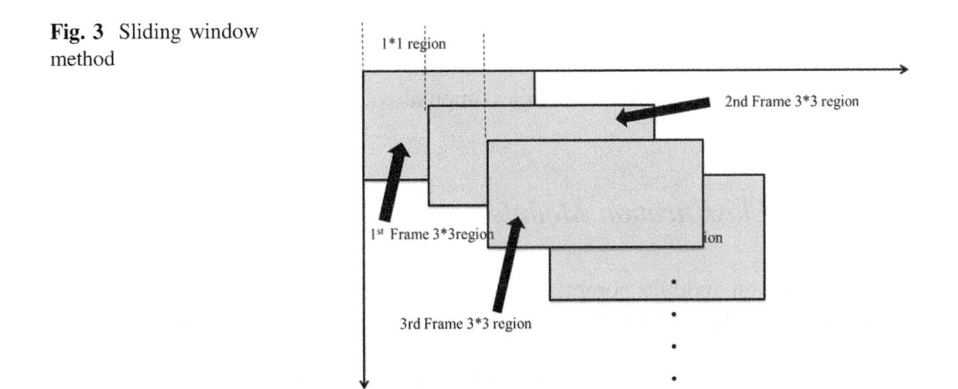

Fig. 3 Sliding window method

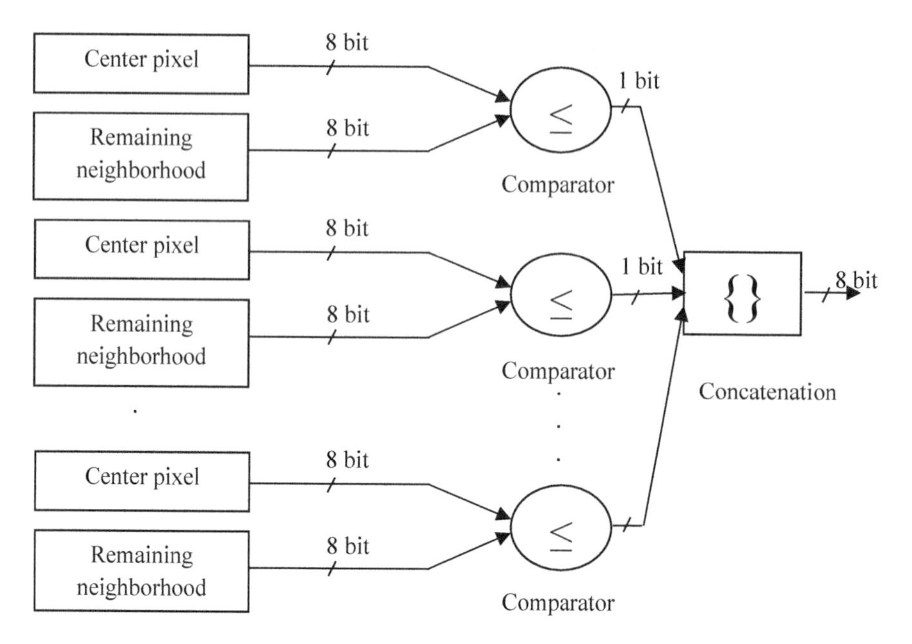

Fig. 4 LBP functional diagram for 3 * 3 frames

3.4 MAX Pooling Based Feature Extraction

The results obtained from the LBP based feature extraction is further used for performing the pool function based feature extraction. The pooling involves the computation of the mean or max value of the particular feature over the region of the image. This function improves the results compared to all other feature extraction process. Here, the mean is computed for particular dimension of an image without overlapping regions for reducing the complexity level. Figure 5 shows the architecture of Max Pooling computation.

3.5 SVM Classification Modules

The classification module comprises of Gaussian kernel unit which involves in computing both the normal processing element as well as exponent processing

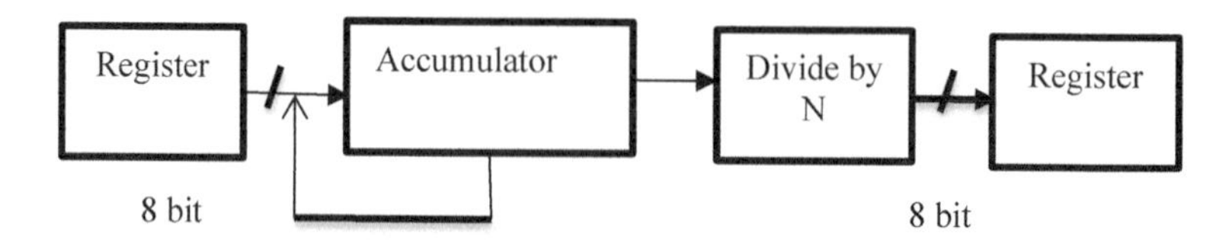

Fig. 5 Architecture diagram for Max pooling computation

element. For the computation of exponent PE, cordic exponent PE unit is preferred. Further decision unit is used to define the subject as drowsy or not.

Gaussian Kernel Unit. The goal of the Gaussian kernel unit is to compute the function $F(x_i, y_j)$. The dot product of the feature space is defined by the kernel function is given by Eq. (3).

$$F(x_i, y_j) = \phi(x_i) \cdot \phi(y_j) \tag{3}$$

The utilization of large number of support vectors provided the satisfactory result in classification process. The computation of kernel function is the difficult task in hardware implementation. To simplify the operation, the VLSI based Gaussian kernel function is defined by Eq. (4)

$$F(x, y) = \exp(-\frac{(x - y)^2}{2\sigma^2}) \tag{4}$$

Here the term σ^2 is the variance of support vectors. The Gaussian Kernel unit contains Processing Elements and serial to parallel units. For each test vector, the unit performs the Kernel evaluations and the respective outcome is sent through the other serial to parallel units. The Gaussian kernel unit contain two modules namely, normal PE and exponent PE. The normal PE architecture is used to calculate the term $A = \frac{(x-y)^2}{2\sigma^2}$ in the kernel function. The test vector is defined by x = (x1, x2 ... x10) and the corresponding support vectors are given by y = (y1, y2 ... y10). The normal PE architecture initially evaluates and estimates the standard deviation and corresponding division sequentially. The architecture diagram for the Gaussian kernel unit is shown in Fig. 6.

Cordic Based Exponent PE. COordinate Rotational Digital Computer (CORDIC) also known as the digit-by-digit method and Volder's algorithm. It is a simple and efficient algorithm to calculate hyperbolic and trigonometric functions. It is commonly used when no hardware multiplier is available (e.g. in simple microcontrollers and FPGAs) as the only operations it requires are addition, subtraction, bit shift and look up table. The exponent function can be implemented by using the Cordic exponent PE unit. The basic block diagram of the exponent function is shown in Fig. 7.

The CORDIC algorithm provides hardware efficient solution since the multipliers are reduced and replaced by adders and shifters which in turn reduces hardware utilization, computational complexity and thereby improves speed. The architecture diagram of the cordic exponent PE is shown in Fig. 8.

Decision Unit. The decision unit is used to define the subject as drowsy or non drowsy. The decision value of each test vector is passed to the decision unit. Based on the threshold, the subject is defined as the output as one or zero. Figure 9 shows the architecture for decision Unit.

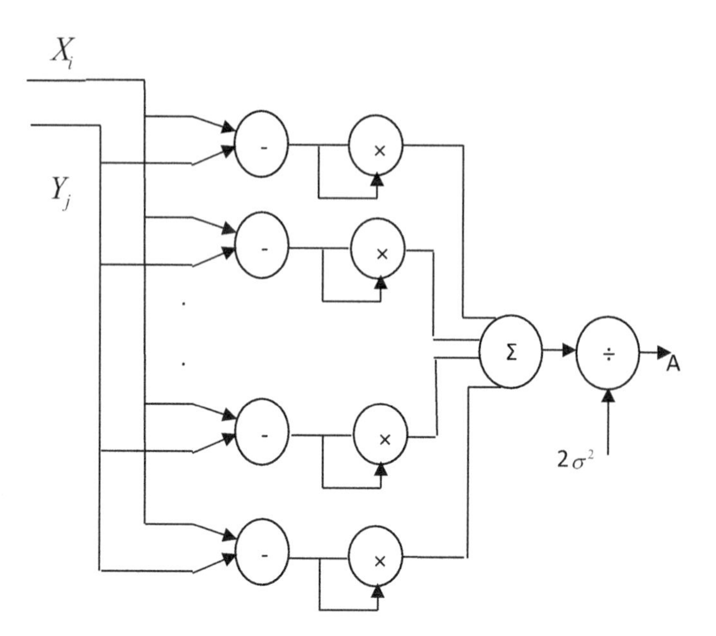

Fig. 6 Architecture diagram for Gaussian kernel unit

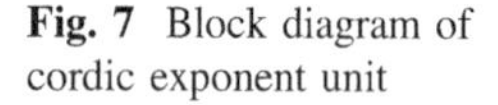
Fig. 7 Block diagram of cordic exponent unit

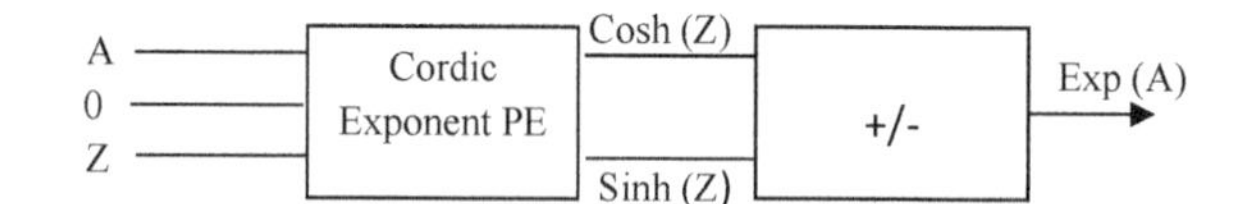

4 Implementation

The implementation involves the hardware as well as software development. The system is based on the Zed Board development board, equipped with a reconfigurable device from the Zync-7000 family. It is a all programmable System on Chip (SoC) which integrates an ARM Cortex-A9 dual core based processor system with Xilinx 7-series FPGA. The overall block diagram of the hardware setup is shown in Fig. 10. The Zed Board which can act as single board computer. The powerful co design platform can be enabled by the Zynq devices, which empowers the user to run the full ubuntu linux desktop. The system is initialized at the startup programming files stored in the SD memory card. The files stored are responsible for the embedded system with specific configuration. Programming files include the Linux boot loader image (U-boot), the device tree for the hardware components hierarchy (devicetree.dtb), zImge and BOOT.BIN. This file contains the information related to the ARM processing system and its configuration. The Linaro Ubuntu desktop image must be stored on the SD card. A separate Linux desktop/laptop machine is needed to program the SD card. The device should preferably have an SD card slot that can be used and accessed by the Linux system.

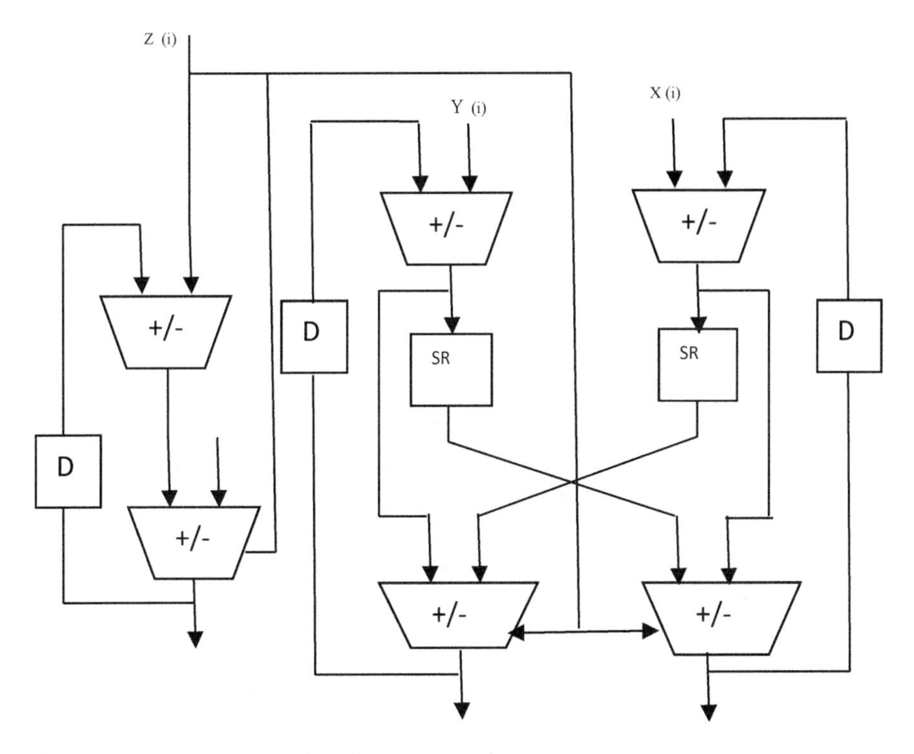

Fig. 8 Architecture diagram of cordic exponent unit

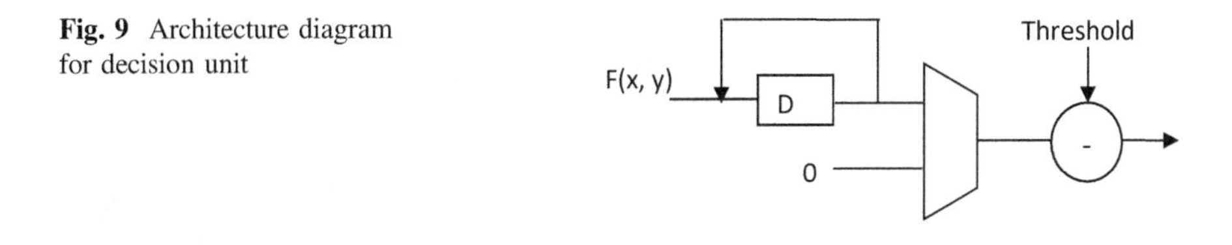

Fig. 9 Architecture diagram for decision unit

4.1 Hardware System

The SD card is positioned into the slot allocated in the device. The USB OTG connector is used to connect the USB powered hub. From the hub connected to the zed board, keyboard, mouse and the camera are connected. The HDMI port is used for connecting the monitor to display.

The hardware setup for the implementation is shown in Fig. 11.

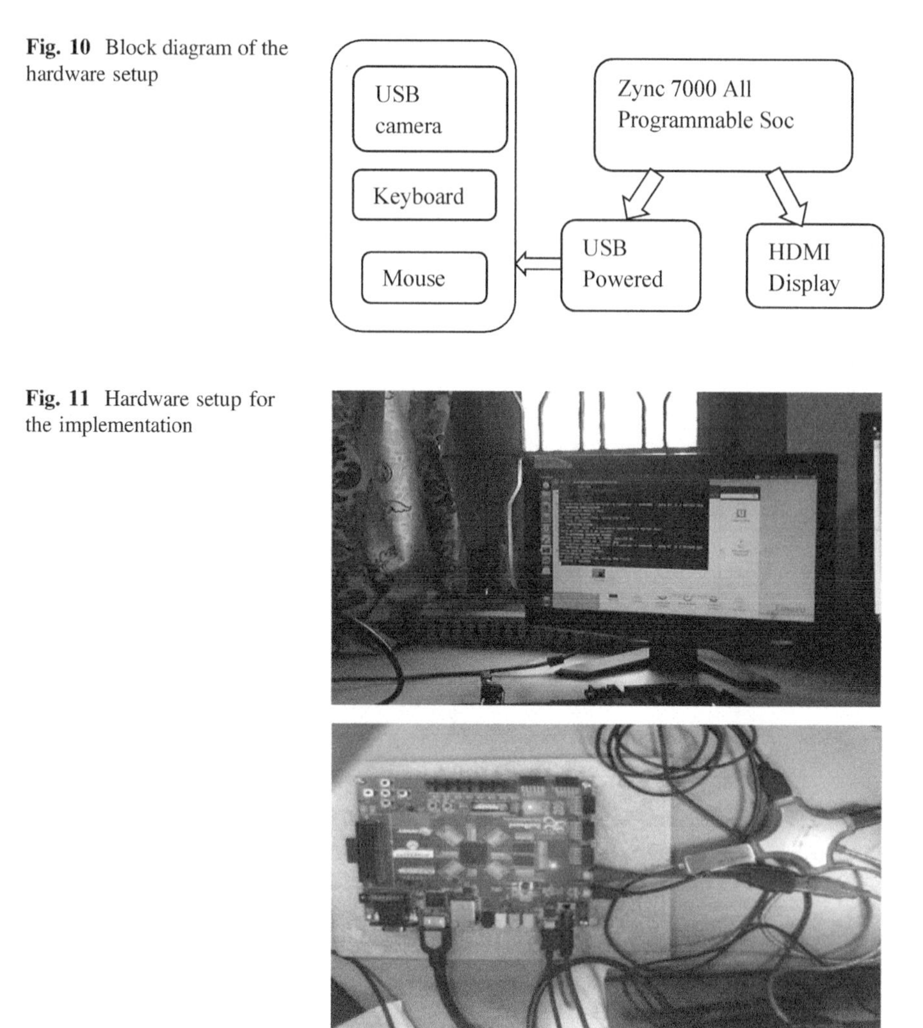

Fig. 10 Block diagram of the hardware setup

Fig. 11 Hardware setup for the implementation

5 Results and Discussion

The VHDL for the proposed method is designed and implemented. The VHDL is simulated using Model Sim Altera 6.4a starter Edition and the implementation is done by using Zedboard—Zync 7000 FPGA using offline images and real time images.

The outputs obtained for some sample drowsy and non drowsy images in the database using preprocessing operations are given in Fig. 12. In preprocessing, the image is subjected to median filtering operation in order to remove the unwanted noise in the image and it is subjected to exactly locate the face using Viola Jones algorithm.

Sample Test images	Input Image	Preprocessed Image
1 Drowsy (offline image)		
2 Drowsy (offline image)		
3 Non Drowsy (offline image)		
4 Non Drowsy (offline image)		
5 Semi Drowsy (real time image)		

Fig. 12 Output results of preprocessing

The output image obtained after preprocessing work is preceded for local binary pattern analysis. The output image obtained is displayed in the VGA monitor by using the FPGA. The image obtained at LBP for sample image 1 and 3 are shown in Fig. 13a, b.

The output obtained at the LBP is given as an input to the Max pooling Layer which results in reducing the complexity level. Figure 14a, b represents the simulation waveform obtained at the output of the Max pooling Function sample image 1 and 3 respectively.

The input vector is fed to the system and the classification of the corresponding simulation output waveform for the test vector having the nature of drowsiness is shown in Fig. 15a, b.

The output image obtained after local binary pattern analysis for real time sample image 5 is shown in Fig. 16.

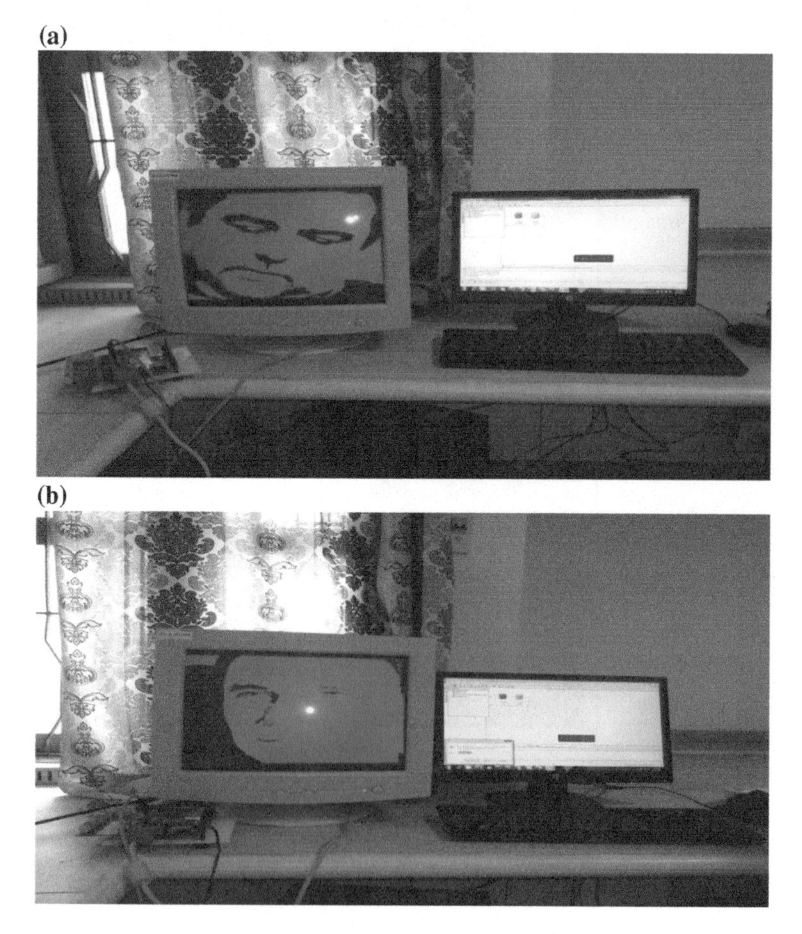

Fig. 13 **a** Output image 1 (drowsy) obtained at the LBP and displayed in the VGA monitor. **b** Output image 3 (non drowsy) obtained at the LBP and displayed in the VGA monitor

Fig. 14 **a** Simulation waveform obtained at the Max pooling function for input image 1 (drowsy). **b** Simulation waveform obtained at the Max pooling function for input image 3 (non drowsy)

The output obtained at the LBP is given as an input to the Max pooling Layer which results in reducing the complexity level. Figure 17 represents the simulation waveform obtained at the output of the Max pooling Function.

The input vector is fed to the system and the classification of the corresponding simulation output waveform for the test vector having the nature of drowsiness is shown in Fig. 18.

From the results obtained using the developed system, it is observed that 96% of accuracy is obtained in detecting the status of the person either as drowsy or non drowsy.

(b)

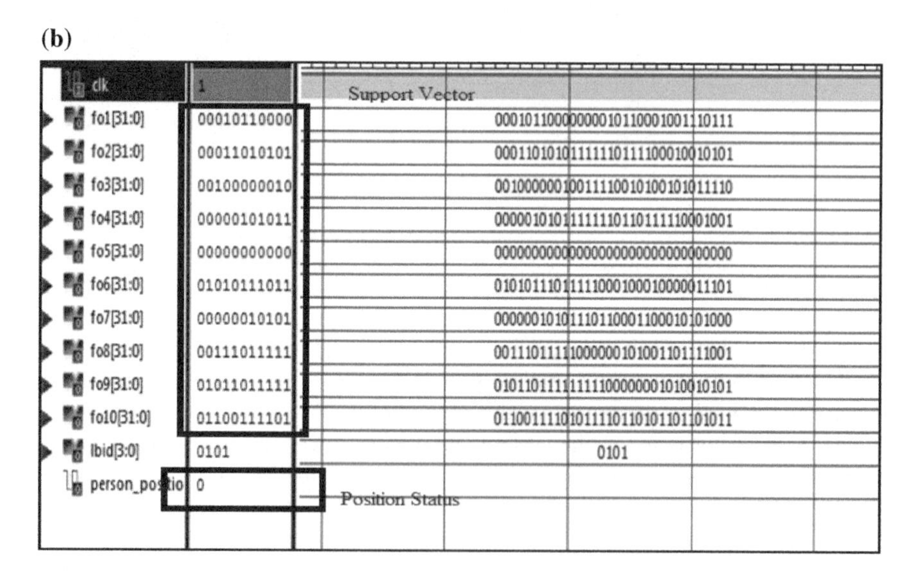

Fig. 15 **a** Output waveform for the person having drowsy (image 1). **b** Output waveform for the person having non drowsy (image 3)

Fig. 16 Output image 5 (semi drowsy) obtained at the LBP

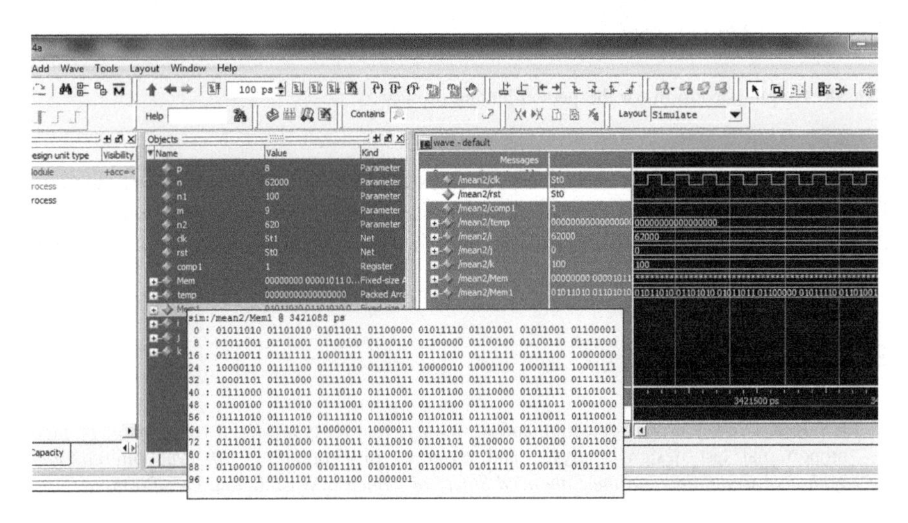

Fig. 17 Simulation waveform obtained at the Max pooling function for input image 5 (drowsy)

6 Conclusion

The driver fatigue and drowsiness detection system is an important criterion for the today's scenario to avoid the causes and some effects occurred due to accidents. Deep learning based feature learning methods are known to provide excellently designed features especially in cases of image or visual data. The LBP based feature extraction with deep learning process is used for the classification. The architecture based classification is preferred in FPGA to reduce the area as well as power consumption. This system is also used for security purpose of a driver. During the monitoring, the system is able to decide if the eyes are opened or closed. When the eyes are closed for long time, a warning signal is issued. Image processing achieves highly accurate and reliable detection of drowsiness. This was achieved by

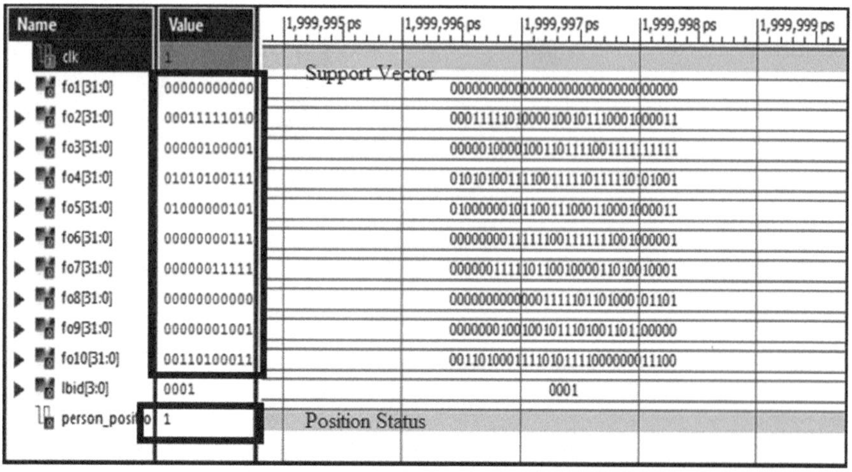

Fig. 18 Output waveform for the person having drowsy

interfacing a webcam to a zed board. The system can find the positions of face and features and further issues the warning. This system is hence used for warning the driver of drowsiness or distractions to prevent traffic accidents.

References

1. P.S. Rau, Drowsy drivers detection and warning system for commercial vehicle drivers: field proportional test design, analysis, and progress, in *Proceedings of 19th International Technical Conference on the Enhanced Safety of Vehicles*, Washington, DC (2005)
2. United States Department of Transportation, *Saving Lives Through Advanced Vehicle Safety Technology*. http://www.its.dot.gov/ivi/docs/AR2001.pdf
3. Y. Takei, Y. Furukawa, Estimate of driver's fatigue through steering motion. IEEE Int. Conf. Syst. Man Cybern. **2**, 1765–1770 (2005)
4. W.A. Cobb, *Recommendations for the Practice of Clinical Neurophysiology* (Elsevier, 1983)
5. H.J. Eoh, M.K. Chung, S.H. Kim, Electroencephalographic study of drowsiness in simulated driving with sleep deprivation. Int. J. Ind. Ergon. **35**(4), 307–320 (2005)
6. A. Samel, H.M. Wegmann, M. Vejvoda, Jet lag and sleepiness in aircrew. J. Sleep Res. **4**, 30–36 (1995)
7. M. Eriksson, N.P. Papanikolopoulos, Eye-tracking for detection of driver fatigue, in IEEE Proceedings of Conference on Intelligent Transportation Systems, pp. 314–319 (1997)
8. X. Zhang, N. Zheng, F. Mu, Y. He, Head pose estimation using isophote features for driver assistance systems, in *Intelligent Vehicles Symposium*, IEEE, pp. 568–572 (2009)
9. Ö. Tunçer, L. Güvenç, F. Coşkun, E. Karsligil, Vision based lane keeping assistance control triggered by a driver inattention monitor, in *IEEE International Conference on Systems Man and Cybernetics (SMC)*, pp. 289–297 (2010)
10. P. Smith, M. Shah, N. da Vitoria Lobo, Determining driver visual attention with one camera. IEEE Trans. Intell. Transp. Syst. **4**(4), 205–218 (2003)

11. A. Liu, Z. Li, L. Wang, Y. Zhao, A practical driver fatigue detection algorithm based on eye state, in Asia Pacific Conference on Postgraduate Research in Microelectronics and Electronics (PrimeAsia), pp. 235–238 (2010)
12. Y.-S. Wu, T.W. Lee, Q.-Z. Wu, H.-S. Liu, An eye state recognition method for drowsiness detection, in *IEEE Conference on Vehicular Technology*, pp. 1–5 (2010)
13. Z. Tian, H. Qin, Real-time driver's eye state detection, in *IEEE International Conference on Vehicular Electronics and Safety*, pp. 285–289 (2005)
14. C.C. Lien, P.R. Lin, Drowsiness recognition using the Least Correlated LBPH, in *International Conference on Intelligent Information Hiding and Multimedia Signal Processing (IIH-MSP)*, pp. 158–161 (2012)
15. A. Lenskiy, J.-S. Lee, Driver's eye blinking detection using novel color and texture segmentation algorithms. Int. J. Control Autom. Syst. **10**(2), 317–327 (2012)
16. T. Pilutti, A. Ulsoy, Identification of driver state for lane-keeping tasks. IEEE Trans. Syst. Man Cybern. Part A Syst. Hum. **29**(5), 486–502 (1999)
17. W. Qiong, Y. Jingyu, R. Mingwu, Z. Yujie, Driver fatigue detection: a survey, in *The Sixth World Congress on Intelligent Control and Automation*, vol. 2, pp. 8587–8591 (2006)
18. A. Picot, S. Charbonnier, A. Caplier, On-line detection of drowsiness using brain and visual information. IEEE Trans. Syst. Man Cybern. Part A Syst. Hum. **9**, 1–12 (2011)
19. K. Hayashi, K. Ishihara, H. Hashimoto, K. Oguri, Individualized drowsiness detection during driving by pulse wave analysis with neural network, in *Proceedings of the 8th International IEEE Conference on Intelligent Transportation Systems*, Austria, vol. 12, pp. 6–12 (2005)
20. R.R. Jhadev, M.H. Godse, S.P. Pawar, P.M. Baskar, Driver drowsiness detection using android bluetooth. Int. J. Innov. Res. Electr. Electron. Instrum. Control Eng. **3** (2015)
21. C. Han, H. Liao, K. Yu, L. Chen, Fast face detection via morphology-based pre-processing, in *Proceedings of Ninth International Conference in Image Analysis and Processing* (1998)
22. Y. Ying, S. Jing, Z. Wei, The monitoring method of driver's fatigue based on neural network, in *International Conference on Mechatronics and Automation*, Harbin (2007)
23. C. Tsai, W. Cheng, J. Taur, C. Tao, Face detection using eigen face and neural network, in *IEEE International Conference on Systems, Man and Cybernetics*, Taipei (2006)
24. D. Liu, P. Sun, Y. Xiao, Y. Yin, Drowsiness detection based on eyelid movement, in *Second International Workshop on Education Technology and Computer Science (ETCS)* (2010)
25. M. Omidyeganeh, A. Javadtalab, S. Shirmohammadi, Intelligent driver drowsiness detection through fusion of yawning and eye closure, in *IEEE International Conference on Virtual Environments Human-Computer Interfaces and Measurement Systems (VECIMS)* (2011)
26. R. Jimenez, F. Prieto, V. Grisales, Detection of the tiredness level of drivers using machine vision techniques, in *Electronics, Robotics and Automotive Mechanics Conference* (2011)
27. K.T. Chui, et al., An accurate ECG-based transportation safety drowsiness detection scheme. IEEE Trans. Ind. Inf. **12**(4), 1438–1452 (2016)
28. G. Li, B.-L. Lee, W.-Y. Chung, Smart watch based wearable EEG system for driver drowsiness detection. IEEE Sens. J. **15**(12), 7169–7180 (2015)
29. F. Rohit, et al., Real-time drowsiness detection using wearable, lightweight brain sensing headbands. IET Intell. Transp. Syst. **11**(5), 255–263 (2017)
30. W.-J. Chang et al., Design and implementation of a drowsiness-fatigue-detection system based on wearable smart glasses to increase road safety. IEEE Trans. Consum. Electron. **64** (4), 461–469 (2018)
31. M. George, R. Zwiggelaar, Comparative study on local binary patterns for mammographic density and risk scoring. J. Imaging **5**(24), 1–19 (2019)

Intelligent Image Analysis System for Position Control of People with Locomotor Disabilities

Marius Popescu and Antoanela Naaji

Abstract The paper presents a remote-controlled system mounted on the seats used by the persons with locomotor disabilities. In these cases, a multitude of problems may be avoided, such as obstacles or bumps, without the direct action of the human being, because the device adapts to each new situation, to reach the destination. The system used to guide or to move a wheelchair on the ground comprises several functional blocks that are distinct as structure, but interdependent. The movement area is monitored by a camcorder connected to the microcontroller, which transmits images to it. The microprocessor processes the images, calculates and sends signals to the communication interface of the equipment. The system receives the commands sent by the microcontroller, interprets them and carries out the movement, together with the transmission of various pieces of information towards the microcontroller such as: confirmations regarding data reception and their validity, data from the sensors, the image itself, etc.

Keywords Image analysis · Avoiding obstacles · Persons with locomotor disabilities · Smart wheelchair

1 Introduction

For adults, independent mobility represents an aspect of self-esteem and it reduces the dependence on attendants and family members, promoting feelings of self-confidence. While many people with disabilities are satisfied with manual or electric wheelchairs, for an increasingly important segment of the disabled community it is difficult or even impossible to use independent wheelchairs.

M. Popescu · A. Naaji (✉)
Department of Engineering and Computer Science,
"Vasile Goldis" Western University of Arad, Arad, Romania
e-mail: anaaji@uvvg.ro

M. Popescu
e-mail: mpopescu@uvvg.ro

D. J. Hemanth (ed.), *Human Behaviour Analysis Using Intelligent Systems*,
Learning and Analytics in Intelligent Systems 6,
https://doi.org/10.1007/978-3-030-35139-7_5

This population includes people with spasticity, tremor, reduced vision or cognitive impairment. These people rely on caregivers to handle their wheelchairs manually.

A wheelchair is built to help an invalid sit down and move to a position in which he can do useful work. In other words, the wheelchair helps people with locomotor disabilities meet their psychological and physiological needs. Many of these people find it difficult to propel wheelchairs with their feet and therefore they opt for electric wheelchairs and especially intelligent wheelchairs.

In order to help this segment of the population and produce intelligent wheelchairs, many researchers have used the technologies developed for mobile robots. Therefore a smart wheelchair consists of either an electric wheelchair to which a computer and a series of sensors are attached, or a mobile robot base attached to a chair.

Intelligent wheelchairs have an on-board computer with which the input sensors can interface, being designed to provide user navigation assistance in different ways, such as providing collision-free transport, performing specific tasks or autonomous moving between locations.

The first intelligent wheelchairs were mobile robots to which sitting parts were attached for people with locomotor disabilities [1]. Subsequently, the development of intelligent wheelchairs for disabled people was based on a change in electric wheelchairs [2, 3]. Most of the smart wheelchair models have been designed as add-on units that can be attached and then removed from the main electric wheelchair [4, 5] in case it gets a malfunction they are mounted on another that is functional.

Worldwide, most wheelchairs are outdated, for severe locomotor disabilities [6] and limiting locomotion is the most frequent disability, making studies that study the effectiveness of robot devices mounted on new and obsolete wheelchairs important. In this way the mobility for young people and adults with disabilities increases and also their physical and social life will be significantly boosted [7].

Electric wheelchairs with intelligent rolling wheels have been the subject of research since the early 1980s [8] as a result of doctors' reports about the use of manual wheelchairs and electric wheelchairs by the patients. Thus, a significant number of patients who cannot use an electric wheelchair have been observed by the medical staff because motor skills or visual acuity are required.

When patients cannot control the wheelchair by conventional methods, it is necessary to have it equipped with an automatic navigation system.

Integrating smart technology into the electric wheelchair features a series of advantages, such as the fact that power up of the processor or seat motors can be bypassed by the manufacturer's control electronics.

There is also the possibility of adding optical encoders to the wheels of the seat, allowing it to track its trajectory.

Systems designed as add-on modules, usually connect to the wheelchair base through interface options provided by the seat manufacturer itself. Thus, the first add-on units [9] used the analog connections between the joystick and the engine controller.

Currently, the add-on modules take over the patented digital technologies in the IT industry, complicating the wheelchair interface very much. For example, in the design of the add-on units, it is necessary to consider the different approaches that have occurred in order to interfere with different wheelchair models.

The advantage of the add-on unit approach is that once the user purchases the intelligent system, he can transfer it to multiple chairs throughout life. This is particularly important for children who grow up over time and therefore use more wheelchairs in a relatively short period of time.

Also, the add-on approach involves more flexible configurations of sensors and input devices, based on the individual needs of each user.

Two intelligent wheelchair models are very known, namely the Collaborative Wheelchair Assistant (which manually controls the direction of a wheelchair that has small powered wheels [10]) and SPAM (prevents collision with obstacles, being compatible with many wheelchair models [11]).

Intelligent wheelchairs have been associated with electric wheelchairs (assembling on them such as joysticks, pneumatic switches, etc.).

Intelligent wheelchairs also used the voice recognition technique [12] as a result of the low cost of implementing the hardware and voice recognition software.

Methods have been implemented for using the view of the machine to calculate the position and orientation of the wheelchair user's head [13]. Voice control was very difficult to implement on standard wheelchairs because the obstacle avoidance capability integrated in the command software needed to protect the patient from the consequences of unrecognized voice commands [14]. To avoid obstacles, intelligent wheelchairs require sensors, and the most used sensor is the sonar (ultrasonic acoustic distance detector). Sonar sensors are very precise, especially when the emitted sound waves reach the object at a straight angle [15, 16]. Another commonly used sensor is the infrared sensor. Infrared sensors emit light, not sound, and they can be confused by light-absorbing materials. Like sonar sensors, the IR ones have difficulty with transparent or refractive surfaces [17]. However, they are very used because they are small and cheap.

However, sonar and infrared sensors are inappropriate for identifying drop-offs, such as curbs, stairs, or ditches encountered on the wheelchair's trajectory. Precise detection of obstacles is possible with laser beam sensors, providing a two-dimensional scanning of 180° in the plane of the obstacles in the wheelchair area. A series of intelligent wheelchairs models use laser beam sensors [18], even though they are expensive, have great sizes and are energy consumers.

Although it was not yet used on a smart wheelchair, another option, cheaper than a laser beam sensor, would be the use of a "laser striper" that consists of a laser emitter and a camera that returns the laser strip image necessary to calculate the distances from obstacles. However, a striper may offer erroneous information when the laser strip falls on a glass or on a dark surface [19]. Because there is not a single, exact, low-cost, small dimension, low-energy, lightweight and waterproof sensor at ambient conditions (illumination, humidity, precipitation, temperature etc.), many intelligent wheelchairs collect information from several sensors to locate obstacles [20]. So the limitations of using a sensor can be compensated for by other sensors.

Therefore, the sonar and infrared sensors are commonly used in combination. From a mechanical point of view, when a smart wheelchair collides with an obstacle, a collision sensor is triggered.

Another technology is the use of photo/video cameras, which are smaller than laser beam sensors and easier to mount in more areas of the wheelchair, also offering a larger coverage area. The cost of the viewing equipment is currently low, and it can be achieved with serial software used for a regular webcam. Intelligent wheelchairs were designed to use a camera and a computer, both for head and eye tracking [21], and for detecting obstacles [22] in order to control the seat.

The approaches regarding the implementing of the intelligent wheelchair control software are based on the functions that the seat must perform and on the sensors that are used. Some smart wheelchairs [23, 24] operate in a similar manner to that of autonomous robots; the patient gives the system a target destination and monitors while the intelligent wheelchair calculates and runs a trajectory to the final location. Usually, to reach the destination, these systems usually require either a complete map of the action area that offers the zone through which they must navigate, or a kind of change of their environment (strip tracks placed on the floor). These types of systems are unable to practice unplanned or unknown areas. Intelligent wheelchairs in this category are suitable for patients who do not have the ability to calculate or follow a trajectory to a given destination, and spend most of their time in the same controlled environment.

Intelligent wheelchairs that use neural networks to reproduce trajectories [25] or use an obstacle density histogram to combine information from sonar sensors with the joystick input from the patient, or using the approaches based on rules (fuzzy) were designed.

Other smart wheelchair models limit themselves only to the collision function, leaving most of the navigation schedules and tasks to the user. The systems from this category do not require prior knowledge of an area or any changes in the environment. They are suitable for patients who can calculate and effectively execute a trajectory to a target destination. There are intelligent wheelchairs that use a multi-layered command architecture where primitive behaviors are connected to develop more complicated behavior [26, 27]. Another category of intelligent wheelchairs offers autonomous and semi-autonomous navigation [28].

Intelligent wheelchairs that offer the semi-autonomous navigation function involve a subset of multiple behaviors, each designed for a specific set of tasks (for example to cross a room avoiding the obstacles, passing through a door, and walking through a hall, beside a wall). Intelligent wheelchairs that provide task-specific behaviors are able to provide a wider range of needs and abilities. The responsibility for selecting the most appropriate mode of operation can be done by the user (manually) or by the intelligent seat (automatic adaptation).

Intelligent wheelchairs that move autonomously to a given destination often generate an internal map. The map can encode the distance (in which case it is referred to as a map value) or it may be limited to specifying connections between locations without distance information (a topological map). There are, of course, other approaches regarding the autonomous navigation, which do not require a system that implies an internal map, such as tracing the tracks on a floor [29].

Intelligent wheelchair models were based on an older technique, meaning that the movement was made by following a local map, memorizing the location of the obstacles in the immediate vicinity of the seat, in order to avoid obstacles [30]. A small number of intelligent wheelchairs are designed to identify the landmarks that appear naturally in the action area. Most intelligent chairs create artificial artifacts that can be easily identified. Thus, most intelligent wheelchairs use the machine software to locate artifacts, or use radio waves.

Modern technologies in the IT and electronics industry will also apply in the future to intelligent wheelchairs. These are ways to test sensors, especially in terms of artificial sight. Smart wheelchairs provide support in analyzing human-robot, adaptive or shared interaction, as well as introducing new methods such as voice control or eye tracking. Furthermore, intelligent wheelchairs will continue to be testing models for robot control architectures. Although there was a major interest in the implementation of intelligent wheelchairs, little attention was paid to assessing their performance, with very few researchers involving people with disabilities in their assessment activities.

Carrying out patient research to develop intelligent wheelchair functions is quite difficult for several reasons. Thus, users do not make any improvement in navigation (in terms of average speed and number of collisions) when using a smart wheelchair on a closed circuit, because the intelligent wheelchair either does not work well or the seat user has little experience, to improve the parameters.

We appreciate the need for long-term studies because the real effects of using a smart wheelchair for a long time are still unknown.

It is possible that the use of a smart wheelchair may reduce the capacity of a patient who relies on the intelligent wheelchair navigation assistance to later use a standard wheelchair. For some users (such as children) the smart technology applied to the wheelchairs is not efficient because it requires a training period and for other users the intelligent wheelchairs will be permanent solutions. Making the difference between using a smart wheelchair as aid for patient mobility as a training tool or evaluation, may in future constitute a study subject for researchers. These functions are unique and require different behavior from every patient using a smart wheelchair.

In terms of mobility, the purpose of the intelligent wheelchair is to help the patient reach a final destination in a fast and comfortable way, by not providing the patient with a feedback to prevent collisions.

In terms of use as a training tool, the goal is to develop specific skills, and feedback will be a function of effective training.

As an assessment tool, the goal of the intelligent wheelchair is to record activity without intervention, with no user feedback regarding the navigation assistance.

Smart wheelchair technology is now used in interior environments that have been modified in order to prevent access to drop-offs.

2 Processing the Images and Color Models

2.1 Problems that Appeared at the Image Acquisition

An image is a description of the variation of a parameter on a surface, but it is not the only parameter used (for example, an image can be generated by the temperature of an integrated circuit). All image types are converted into classic images by pseudo-coloring, so that the human eye can make a visual evaluation of the variation of some parameters. In the classic sense, a picture is a two-dimensional signal, and the image processing is considered a branch of digital signal processing.

Image processing and, in general, digital signal processing imply a relatively large use of computing and memory resources. Therefore, the implementation of specific algorithms can be done on classical computer systems (Personal Computer, PC), and for dedicated real-time systems, dedicated processors commonly called digital signal processors [31] are generally used. Image processing involves the accumulation of a certain amount of information (base knowledge, environmental information, etc.) that is taken over by the intelligent sensor systems and creates an image of the environment at the time of the takeover of the data (snapshot). From the image thus obtained useful information must be extracted.

A digital image acquisition system consists of a series of devices: an optical system (lenses, a diaphragm), a sensor (CCD—Charge Coupled Device), a potential amplification and filtering step of the sensor signal (the information is still analogous) and an analog-to-digital converter. The image acquisition system is an essential component of the image processing chain, and in its design it must be taken into account a multitude of optical parameters (lens type, focal distance, field depth), photometric (type, direction and intensity of illumination, reflection properties of the analyzed objects, the characteristic of the photo-receiver output) and geometric (projection type, perspective distortions, position and orientation of the sensor). The acquisition system will produce a digital image, which is a two-dimensional array whose values can represent the intensity of light, distances or other physical sizes.

The images were classified in four classes, respectively, Class 1, which includes color or grayscale images (television, photography), Class 2, which corresponds to binary (two-color) images, Class 3, comprising images consisting of lines and continuous curves, and Class 4, which corresponds to images composed of isolated points and polygons.

This classification also takes into account the complexity of the images. With the class number it also decreases the complexity of the image and, implicitly, the amount of data needed to store it (for example, a Class 1 image may be the image of any object-photo). Following a binarization operation, a Class 2 image will result. By applying contour extraction algorithms or morphological operators (dilation, erosion), a Class 3 image is obtained. Finally, by extracting the critical points, the areas of interest a picture of the last class will be created.

2.2 Graphic Processing

Preprocessing operations that take place on captured images either have the role of eliminating unnecessary noise or useless information from the image, or they are restoration operations. Such processing is necessary to improve both the execution times and the results of various algorithms (classification, recognition forms, human face recognition etc.).

The filtering operation is used to eliminate noise and for highlight the edges. Since the images are two-dimensional signals, which can be decomposed into sums of sinusoidal signals, the notion of filtering and spectral analysis, valid for one-dimensional signals, are also valid for two-dimensional images (obviously with some modifications due to the nature of the signals).

The image restoration operation is required as the output from the image acquisition block results in a distorted image, distortions that occur due to known physical phenomena. A possible source of distortion is the optical system, and the purchased images may show pillow, barrel or trapezoidal distortions. These errors can be corrected by a resembling operation (for these types of geometric distortions the mathematical relationships are necessary for the correction can be determined, and, using these relationships, the value of each sample/pixel in the new image is calculated based on a number of samples from the distorted image). Also, because of the optical system, the images may have different illuminations on portions. A luminous point corresponds to a sensor circle on the sensor surface, called a circle of confusion (due to the lens). For normal objectives, this circle is uniform.

There are manufacturers of optical systems which, for various reasons (apparent increase in resolution, etc.) produce under-corrected objectives (the circle is brighter in the center) or over-corrected (the circle is brighter on the edge).

Segmentation is a process of dividing the digitized image into subsets by assigning individual pixels to these subsets (also called classes), retrieving distinct objects from the scene. Segment algorithms are generally based on two principles: discontinuity (with the main method of contour detection) and the similarity (with the threshold method and region method). An image's histogram is a function that indicates how many pixels (a pixel is represented on 1 byte) have a certain gray level. In an analysis of the histogram of the various types of images, it is usually found that a picture has several levels of gray and has two local peaks. Using this feature, a threshold segmentation of the original image can be made, choosing the threshold as the local minimum between the two maximum points. Thus, the pixels with a gray level smaller than the chosen threshold can be considered as, for example, the background (they are assigned the value corresponding to the black), and those for whom the gray level is higher than the threshold are the pixels of the object. Because of poor, uneven illumination, or around the sensitivity threshold of the CCD sensor, low contrast images may result, images for which pixels do not take values across the entire range of gray levels. To correct these defects, histogram alignment or linear extension is usually used across the entire range of gray levels. Through these operations, a pixel in the new image is assigned a gray value calculated based on a function (linear or exponential), functions that are determined based on the original image. To extract the contour there are several operators, the most known being the three classical operators—of the type of sliding window: Sobel, Kirsch and pseudo-Laplace [32]. Extracting the contour through these operators consists of a sequence of convolutions between the initial image and the operator's nucleus (mask). Optionally, for each operator, at the end of the convolutions, threshold segmentation can be made to get a binary image of the edge map. Other operators used to extract/detect edges are Deriche and Canny [33]. Identifying objects in an image is difficult to achieve, and one method is to binarize gray-scale images and then apply morphological operators. The vast majority of algorithms in this category are simple operations. The most frequently used morphological operators are also the eroding and dilation, for which it is considered a binary image containing the object of interest and the background (which does not interest), for example the background is white and the object is black. By eroding each pixel of the object that touches the background is converted into background pixel. At dilation, each pixel in the background that is in contact with the object is converted into object—pixel. Thus, erosion makes objects smaller and can divide a larger object into several smaller objects; dilation makes objects larger and can unite more objects.

On the basis of these operators, there is an opening operation (an erosion followed by a dilatation) and closure (dilation followed by erosion). The first operation is used to remove from the image the small "islands" from the object pixels that usually appear to binarization because of inadequate threshold selection. Similarly the closure operation eliminates such "islands" from the background. These techniques are useful in processing noisy images where part of the pixels may have an erroneous binary value (for example, an object may have no holes or the edges of the object will not show any roughness). Motion detection and tracking of various moving objects is an operation commonly used in video surveillance systems. There is generally no generic approach, but the algorithms or solutions found are usually unique to the problem. Mainly, a comparison/difference between successive frames and/or between the current frame and a reference frame is made.

One of the issues that occur is the separation of the moving object from the rest of the scene (generically called background). A possible solution would be calculating the difference between frames, threshold segmentation, noise filtering, or creating a motion image. Another approach would be to use segmentation by the region method to identify contiguous blocks of motion that are found in the difference between two successive frames. The absolute value of the difference between two successive frames is submitted to a threshold operation resulting in a rough image of the motion. From this, a motion map is obtained in which each cell (pixel) corresponds to a 3×3 window from the previous image, for which each pixel in the window is above a threshold. This image size reduction offers greater noise tolerance and reduces processing time. From the new image of the movement thus obtained, the largest contiguous movement block is chosen.

Form recognition is a commonly used way of extracting information from the purchased images. It is a wide domain, including human face recognition, fingerprint recognition, etc. Form recognition consists of a classification and/or a description of the image content. Classification consists in assigning an unknown shape from the image taken to a class from a predefined set of classes. The classification operation will output a new image, representing a map of the objects in the scene. In the new image, the pixel values are actually the codes associated with the corresponding classes. Classification uses mathematical methods called theoretical-decision-making or statistical methods of recognition that are based on elements from the statistical decision theory. Classification algorithms are based in the extraction of features on a measure of similarity (for example, a distance). An important step in the design of automatic classification systems is the selection of features, because the components of the feature vector imply the presence of a rather large amount of information. This selection is a problem dependent on the number of classes and the analyzed forms.

2.3 Color Models

Equipment-oriented [34] is based on primary colors used by color rendering equipment; in this category are RGB (Red, Green, Blue), CMY (Cyan-Magenta-Yellow) and YIQ. User-oriented are based on the psycho-physiological properties of the colors: HSV (Hue, Saturation, Value) and HLS (Hue, Lightness, Saturation). A color pattern specifies a 3D coordinate system and a subspace of colors in the respective coordinate system. Each color is represented by a point in the color subspace.

The RGB model is used to represent colors in light transmitting equipment. The color subspace is the unit cube, defined in the Cartesian 3D coordinate system (Fig. 1a) [35]. Each axis corresponds to a primary color: red, green, and blue. A color is specified by three real values between 0 and 1, representing the contributions of the three primary colors to color formation. The gray shades are represented by the main diagonal points and correspond to the mixture in equal proportions of the three primary colors.

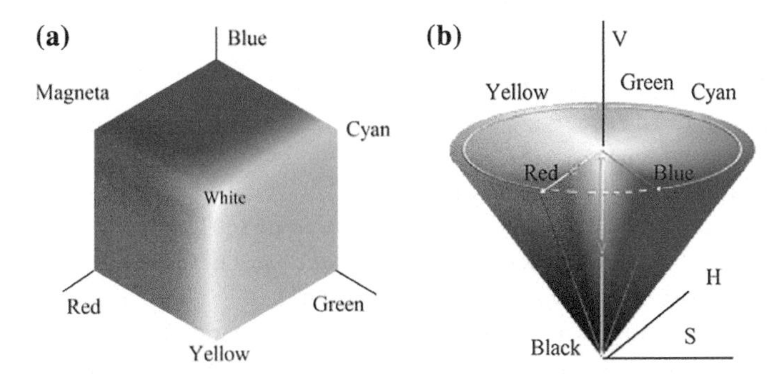

Fig. 1 Color models: **a** RGB model: Red (1, 0, 0), Green (0, 1, 0), Blue (0, 0, 1), White (1, 1, 1), Cyan (0, 1, 1); **b** HSV model

Specifying colors in RGB and CMY models can be difficult for the users of a graphical system. Artists specify colors through tints, shades and tones. Starting from a pure pigment, an artist adds white to get a tint, black to get a tint, white and black to get a color tone. By placing the corresponding triangles to pure colors around a central black and white axis, a three-dimensional hexagon representation is obtained (Fig. 1b). The hexagon is the color subspace in the HSV model. The HSV model uses the hue, saturation, and value concepts with the following parameters:

$$0 <= V <= 1,\ 0 <= S <= 1,\ 0 <= H <= 360 \tag{1}$$

The hexagon base corresponds to the maximum intensity colors (V = 1).

The hue is defined by the angle of rotation around the vertical axis, the zero angle corresponding to the red color. The complementary colors are situated 180° apart from each other on the base of the hexagon. Saturation is defined by the distance from the hexagon axis at the edges of the hexagon for all V values. It varies from zero, on the axis, to 1, on the sides of the hexagon. A combination of only two primary colors gives a maximum saturation color, but a mixture in which all three primary colors are different from zero cannot produce a color with maximum saturation. The combination (S = 0, V = 1) corresponds to the white color, and by (S = 0, 0 <= V <= 1) the gray levels are represented. For S = 0 the value of H is insignificant. Pure colors are represented by V = 1 and S = 1. Selecting the pure pigment with which is chosen to start is the choice of H. Adding white means reducing the S. Adding black means decreasing the V. Different tones are obtained by decreasing both V and S. The hexagon base corresponds to the surface seen when it comes to the RGB cube from the tip (1, 1, 1) along the main diagonal. Each constant V plane in the HSV space corresponds to the view of another cube from the RGB cube.

The main diagonal of the RGB cube becomes the main axis in the HSV space. This interpretation is the base of the conversion algorithms between RGB and HSV models, such as the one played in the following program sequence [36]:

```
float r,g,b,h,s,v;int RGB_HSV()
{/*We give r, g, b, values between 0 and 1 then calculate s,v be-
tween 0 and 1, h between
0 and 360 degrees; the function returns 0 if h is
undefined, 1 if calculated*/
float max,min,dr,dg,db;max=maxim(r,g,b);min= minim(r,g,b);
/*brightness calculation*/
v=max;/*saturation calculation*/
if (max!=0) s=(max-min)/max;else s=0;/*hue calculation*/
if(s!=0) {dr=(max-r)/(max-min);/*approaching the color to red*/
dg=(max-g)/(max-min);db=(max-b)/(max-min);
   if(r==max)/*The color is between yellow and magenta*/
   h=db-dg; else if(g==max)/* The color is between cyan and yellow*/
h=2+dr-db; else /*b==max*/h=4+dg-dr;h=h*60;
/* The color is between magenta-cyan, conversion in degrees*/
```

```
if(h<0) h=h+360; return 1;}else return 0;}
int HSV_RGB(){/*h,s,v are given; h is between 0 and 360 degrees or it
is undefined (<0) s and v are values between 0 and 1*/
 int i; float f,p,q,t; if(s==0) /*achromatic light*/
  { if(h<0) /*h is undefined*/
  {r=g=b=v; return 1;} else return 0;}
 if(h==360) h=0; h=h/60; i=h; /*the integer of h*/
 f=h-i; p=v*(1-s);q=v*(1-(s*f));t=v*(1-(s*(1-f)));
 switch(i) {case 0:r=v;g=t;b=p; break;
case 1:r=q;g=v;b=p; break;
case 2:r=p;g=v;b=t; break;
case 3:r=p;g=q;b=v; break;
case 4:r=t;g=p;b=v; break;
case 5:r=v;g=p;b=q; break; }return 1;}
```

There are several methods used to illuminate 3D scenes based on color interpolation. For instance, calculating the intensity in the Gouraud model, calculating the illumination at a point on a transparent surface, and others [37]. The result of interpolating between two colors depends on the color pattern in which they are specified. The result of interpolation in any of the RGB, CMY and YIQ models will be the same. The result of interpolation in HSV or HLS will be different from interpolation in RGB. In the program sequence below is shown the mode of generation of the HSV model:

```
var i,j:integer;RGBTr:TRGBTriple;
begin for i:=0 to Image1.Width do
  begin RGBTr:=HSVtoRGBTriple(i,255,255);Image1.Canvas.Pen.Color:
=RGB
(RGBTr.rgbtRed,RGBTr.rgbtGreen,RGBTr.rgbtBlue);Image1.Canvas.
MoveTo(i
,0);Image1.Canvas.LineTo(i,Image1.Height-3);
    RGBTr:=HSVtoRGBTriple(Hue_obiect,255,255);
    j:=RGB(RGBTr.rgbtRed,RGBTr.rgbtGreen,RGBTr.rgbtBlue);
    if toleranta then
      if ((abs(i-Hue_obiect)>Hue_toleranta) and
         (abs(i-Hue_obiect)<(360-Hue_toleranta)))
then j:=RGB(255,255,255);
    if (i=Hue_obiect)then Image1.Canvas.Pen.Color:=clBlack
    else Image1.Canvas.Pen.Color:=j;
    Image1.Canvas.MoveTo(i,Image1.Height-3);
    Image1.Canvas.LineTo(i,Image1.Height);
  end; end;
```

3 The Add-on Mode for the Electric Wheelchairs

3.1 Hardware Description

Electric wheelchairs with intelligent wheels are important solutions for people with locomotor disabilities who cannot carry out their daily activities using only motorized electric seats. The device used to control the electric seat, called an access device or control mechanism (which can be operated using various types of joysticks or switches), can be used with medium control systems or computer access. The intelligent wheelchair that will be presented below is provided with several functional blocks distinct as a structure but interdependent. The indoor environment is monitored by a video camera that is connected to a specialized computing system (tablet, smartphone, computer, etc.) which transmits images to it.

The microprocessor processes the images, calculates and sends signals through the serial port to the add-on mode interface. It receives the commands sent by the microprocessor, interprets them and carries out the movement, together with the transmission to the computing system of various information such as confirmations received by the wheelchair data and their validity, data from sensors etc.

The ultimate goal is to move the intelligent wheelchair to an arbitrarily chosen point in the image captured from the camcorder after a calculated and automatically generated trajectory by the computing system. At the same time, necessary corrections are made to deviations from the calculated trajectory of the robot, due to the unevenness of the area of action or the intervention of the various external factors. Communication with the patient consists of an interface that connects the smartphone, the command transmission and data reception from the proposed device, and the emission-reception modules.

The assembly of the mechanisms (motors, couplers) is coupled through an interface which connects the train motors and the mode of reception of the commands and the data transmission, respectively, the complementary modules for receiving/transmitting to the communication interface (Fig. 2).

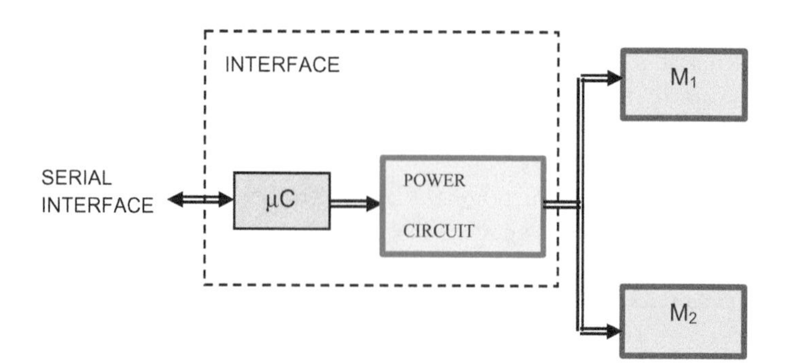

Fig. 2 Block diagram of the add-on mode

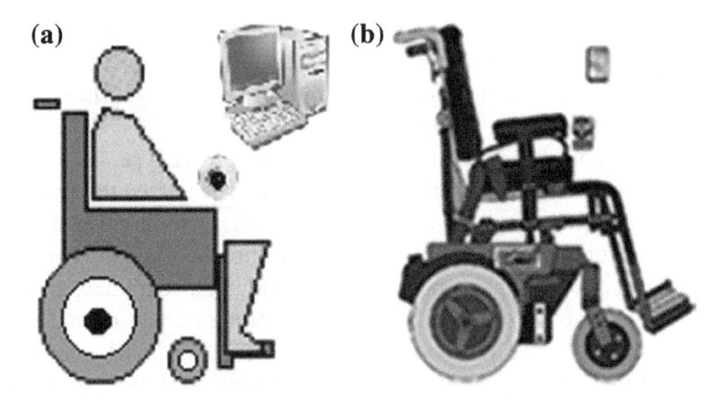

Fig. 3 Attaching the camcorder and add-on mode to an electric wheelchair: **a** personal computer/tablet—Windows/Linux; **b** smartphone—Android/iOS

Choosing the camera connected to the device depends on the environment and the conditions in which the wheelchair patient lives. So, for maximum efficiency, the action area must have a single background color to avoid creating confusion when processing images. By means of the video surveillance camera, images are taken from the area where the wheelchair is located, images representing the information according to which the order is given, which is also the reaction information. The camcorder is mounted on a support above the action area so that the entire area is visible in order to observe all possible positions of the wheelchair (Fig. 3).

The following program sequence shows the connection mode of the camcorder:

```
procedure TForm1.B_ConectareClick(Sender:TObject);
begin if CPDriver.Connected then CPDriver.Disconnect
        else CPDriver.Connect;
case CPDriver.Connected of
true:begin B_Conectare.Caption:='Deconectare COM';
       B_setariCOM.Enabled:=false; end;
else begin B_Conectare.Caption:='Conectare COM';
       B_setariCOM.Enabled:=true;end;
end; end;
```

Thus, the wheelchair becomes an intelligent one, that is, a robot that can move in a plane. A square-shaped plate was drilled in the central part and recesses were made for mounting a bearing. On this plate it was mounted a disc that provided a clamping system so that it is movable relatively to the plate.

Between the disc and the plate, a pressure bearing was mounted which, due to its shape, allows the clamping system to tighten the disc without blocking its movement towards the plate. The wheel drive is achieved by direct contact with the axle of the gearbox, both the wheel and the reducer having rubber rings. The reducer is built in an open version without a protective casing and consists of two dented wheels which make a transformation ration of 22/1.

The assembly of the continuous current motor, reducer, transmission and the wheelchair's wheel is shown in Fig. 4.

The electric motor of the reducer is a continuous power type.

Two metallic concentric rings are mounted on the disc to feed the engine used to drive the reducer. These rings are in permanent contact with a brush system, fixed on the robot structure, to make the electrical connection between the engine and the electronic control circuits. In the robot module structure, a continuous power motor was mounted on the base plate, driving the disc through a belt having the role of directing the seat [25]. The robot-mode brain is the AT89C2051 microcontroller, produced by Atmel, a company that offers 8/18 bit microcontrollers based on the C2051 architecture. This programming device was designed to be flexible, economical and easy to build; the hardware of the programming device uses the particular TTL (Transistor Transistor Logic) series and no special components are used. The programmer is connected to the parallel port of the computing system and there are no special requirements for the parallel port, so older computers can also be used. Any computer system and any operating system (Windows, Linux, Android, iOS-Apple) can be used and it requires no special software except a simulator program. The electronic circuit diagram of the robot module is comprised of the AT89C2051 integrated circuit (Fig. 5).

This circuit manages the add-on module control with the required auxiliary circuit elements—the external oscillator circuit (quartz) and the RESET circuit—and the force circuit L293, required to power the robot-mode motors. The external oscillator is made with 12 MHz quartz and it gives the microcontroller's frequency. The P1 port lines were used to control the L293 integrated circuit which is a driver capable of distributing current at a 1 A output to the channel. Each channel is commanded by a TTL logic gate. A separate voltage source supporting the logic gate is used to operate at low voltage to reduce the losses. The L293 circuit has 16 pins and uses the four central pins to drive the heat to the cooling radiator. This integrated circuit supports a supply voltage for the connected motors, V_S, up to 36 V, and as voltage for logic levels, V_{SS}, up to 36 V. The maximum voltage

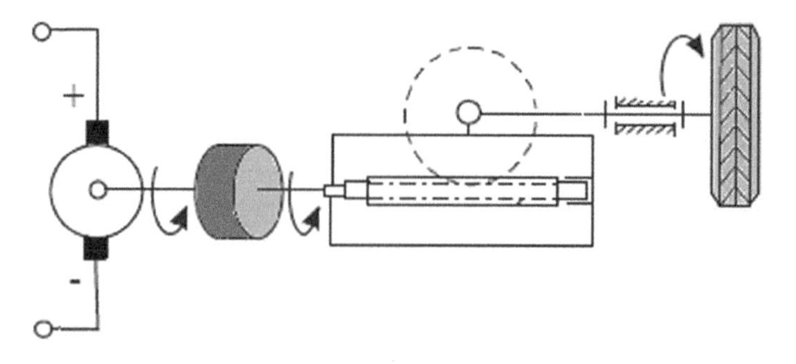

Fig. 4 The continuous current motor, reducer, transmission and wheel of the chair assembly

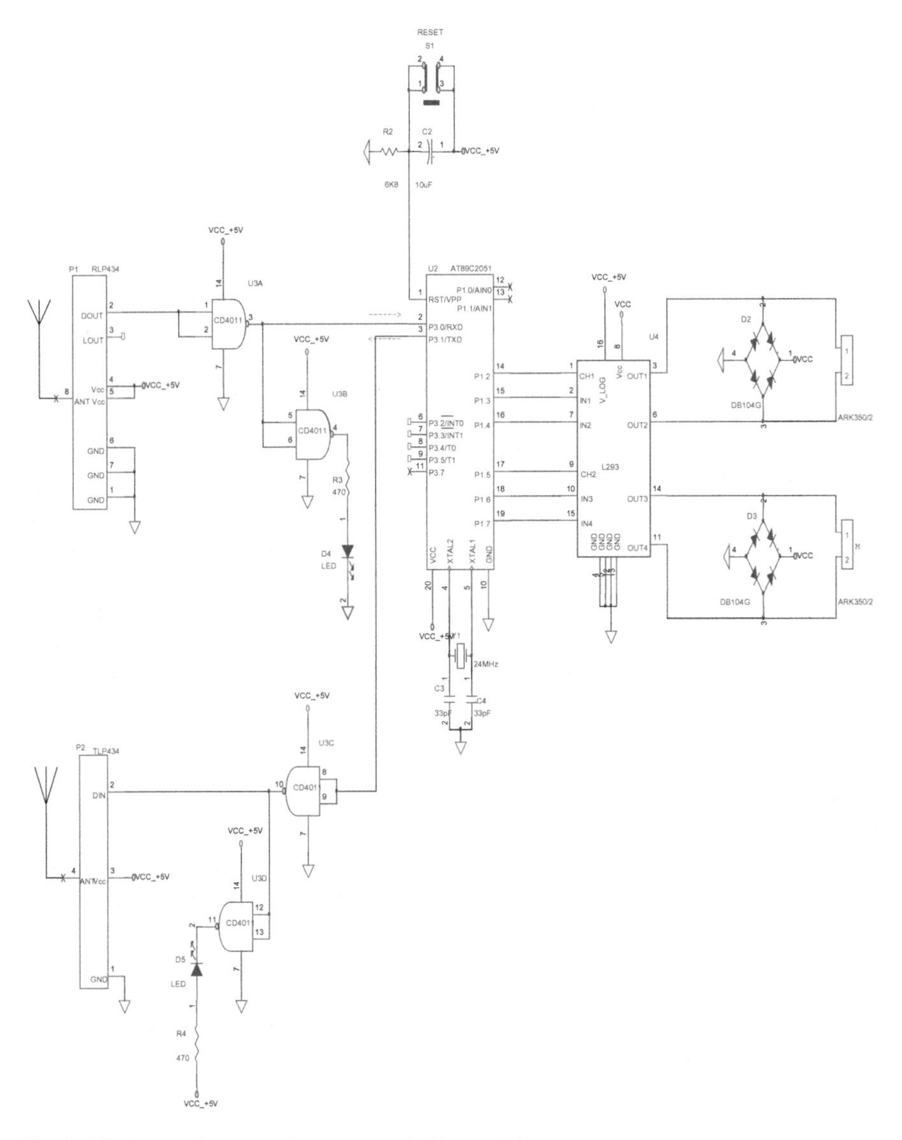

Fig. 5 Electronic diagram of the microcontroller module

accepted for an input, V_I, is max. 7 V. In the electronic diagram used to command the robot module, two bridges of diodes were used for each motor instead of the limitation diodes, that were mounted in such a way as to have the same effect, but the number of paths from the wiring were reduced.

It is specified that the module can be programmed to perform certain movements without external supervision or it may have a connection to a computing system, for example, indicating the commands it has to carry out. A very simple way to connect to a microcontroller is to use the serial lines of the microcontroller. These lines can be connected directly to the serial port of the computer system (PC), either through wires or radio waves. From a security point of view, it is recommended to use the 802.11i security standard and the WPA2 (Wi-Fi Protected Access II) diagram. Choosing other simple access control techniques to a 802.11 network is considered unsafe, such as the Wired Equivalent Privacy (WEP) scheme, dependent on an unreliable symmetric encryption algorithm, RC4 [38].

It is also possible to implement a Bluetooth connection, which is stable, with low energy consumption, but also with great limitations regarding distance.

In addition, the infrared module has, at its turn, a number of advantages: low cost, simplicity, low energy consumption; but also has a number of disadvantages, which in this case are indispensable, namely, the coverage is very small. So, the Wi-Fi mode is the most efficient, although it consumes more power than the other variants, but the drive distance is great. Besides, it can be accessed by any device with a modern browser and an internet connection. In order for the robot module to be independent (without any external physical connection—cables, wires), a radio communication between the robot module and the computing system was implemented (Fig. 6).

For radio transmission and reception, transmission and reception modules produced by Laipac, respectively TLP315, RLP315, TLP433 and RLP433, operating at the frequency of 315 and 433.92 MHz, were used. The TLP emission circuit can be powered at a tension between 2 and 12 V, the maximum current absorbed at the 2 V voltage supply is 1.64 mA, and the 12 V supply voltage is 19.4 mA. The RLP (the circuit used for reception) can be powered at a voltage between 3.3 and 6 V.

These modules work using amplitude modulation and allow a transfer rate of between 512 bps and 200 kbps. A big advantage of these modules is that they allow digital transmission respectively reception, without the need for analogue-to-digital or digital-analog conversions. Unlike analog transmission, digital transmission has high noise immunity, requires a small number of communication circuits. Thus, communication between two numerical devices can be made simply by using these modules, the only condition being to be able to work serially.

The modules used are of a small size, without the need for circuits other than for the adjustment of logical levels.

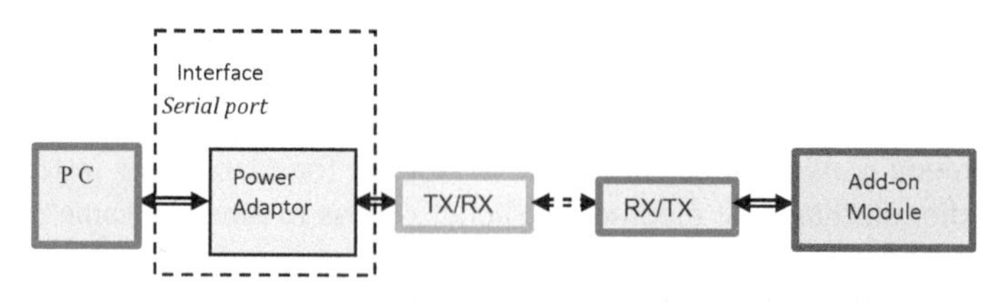

Fig. 6 The electronic-block diagram of the communication interface

For the interfacing of the transmitting and receiving radio modules with the computing system, the MAX232 integrated circuit was used which is a level converter with Charge-Pump specialized circuits. It converts the logical levels received from the input in the corresponding levels for the circuits at output meaning: it converts the logical levels ±12 V received from the serial port of the computer on the TxD line, in the logical levels TTL of 0 ÷ 5 V required for the majority integrated numerical circuits; converts the logical levels of 0 ÷ 5 V received from the integrated numerical circuits from the exterior in the logical levels of ±12 V necessary for the serial port of the computer on the RxD line.

In order to obtain impulses with an unattended form, an integrated circuit of CD4011 buffer role, was used between the MAX232 circuit and the transmission-receive modules. This circuit has four NAND logic gates incorporated.

Led lights were mounted for signaling the fact that it is being emitted or received on the lines.

At the robot module circuits, the radio transmitting/receiving side has been boosted so that communication can take place properly.

3.2 *The Conduction Algorithm and Operating Mode*

In order to automatically drive the wheelchair in robot mode, having a smartphone with a video camera and a communication interface, an algorithm whose protocol is described below was implemented.

The camcorder captures images from the area where the wheelchair is fitted with the robotic device. The image is perceived on the computer as an array whose size depends on the resolution of the camcorder. Each element of the array corresponds to one pixel in the image and consists of four bytes, the first three bytes being the red, green, blue values of that pixel.

Once the image is captured, it is subjected to a transformation that facilitates further processing. In order to coordinate the movement of the wheelchair in the entire area monitored by the video camera, it is first necessary to detect the position of the object in the captured image.

Two methods can be used for this. The first uses a threshold value (generally applies in situations where the background color is uniform and very different from the object's, in order to make the separation of the background object as good as possible) and the second method uses the RGB transformation in HSV.

This is done as follows: from the values R, G and B, the maximum value is set and the Max is marked, and the Min is the minimum of these values.

According to the formulas below, the values H, S, and V result [39]:

$$H = \begin{cases} 60 \times \frac{G-B}{Max-Min} + 0, & \text{if Max} = \text{R and G} \geq \text{B} \\ 60 \times \frac{G-B}{Max-Min} + 360, & \text{if Max} = \text{R and G} < \text{B} \\ 60 \times \frac{B-R}{Max-Min} + 120, & \text{if Max} = \text{G} \\ 60 \times \frac{R-G}{Max-Min} + 240, & \text{if Max} = \text{B} \end{cases} \quad (2)$$

$$S = \frac{Max - Min}{Max}, \quad V = Max \quad (3)$$

The color hue (H) varies between 0 and 359, indicating the angle in degrees in the color circle where the hue is located. The color saturation (S) has values ranging from 0 to 255 and the higher it is, the stronger the color, for smaller values, the color tends to white. The value of the color (V) has values between 0 and 255 and the higher it is, the purer the color is and for smaller values the color approaches black. For example, for S = 0 and V variable grayscale is obtained.

This method has been used to identify the robot device in various illumination situations, so the program can control the device and when the work area is illuminated more strongly also when it is illuminated weaker.

The color distribution mode in the HSV model (hue, saturation, value) is a cylindrical geometry, with hue, angular dimension ranging from red to 0°, passing through green to 120° and blue to 240° and then wrapping back red at 360°.

In the program sequence below, the image generation mode is displayed:

```
GenereazaImagine(self);
if (Button1.Tag=1) then Timer1.Enabled:=true;end;
procedure TForm1.GetActualPosition(Sender:TObject);
var x,y,x_suma,y_suma,ch,cs,cv:integer;SL:PRGBArray;
beginVC.OnFrameCallback:=VCFrameCallback;
VC.GrabFrameNoStop;VC.OnFrameCallback:=nil;
if (VC.FrameData<>nil) then
  VC.FrameToBitmap(bmpV,VC.FrameData.lpData,VC.BitMapInfo);
x_suma:=0;y_suma:=0;nr_pct:=0;
for y:=0 to h-1 dobegin SL:=bmpV.ScanLine[y];
    for x:=0 to w-1 dobeginRGBTripleToHSV(SL[x],ch,cs,cv);
if (((abs(ch-Hue_obiect)<Hue_toleranta) or
           (abs(ch-Hue_obiect)>(360-Hue_toleranta))) and
            (cs>20) and (cv>20) //it is colored
) then begin nr_pct:=nr_pct+1;
x_suma:=x_suma+ x; y_suma:=y_suma+ y;
if CB_FB.Checked then begin
SL[x].rgbtRed:=SL[x].rgbtRed+5*((255-SL[x].rgbtRed) div 6);
     SL[x].rgbtGreen:=SL[x].rgbtGreen+5*((255-SL[x].
rgbtGreen) div 6);
SL[x].rgbtBlue:=SL[x].rgbtBlue+5*((255-SL[x].
```

```
rgbtBlue) div 6); end; end elsebegin
SL[x].rgbtRed:=255;SL[x].rgbtGreen:=255; SL[x].rgbtBlue:=255; end;
   if nr_pct>0 thenbegin//poz_act.x:=x_suma div nr_pct;
//poz_act.y:=y_suma div nr_pct; end; end;end;
```

In each geometry, the central vertical axis comprises neutral, achromatic or gray colors ranging from black to brightness 0 or value 0, the bottom, white to brightness 1 or value 1, top.

The image thus processed, is scanned pixel with pixel and it only those pixels that have the same color as the one selected or resulting from object detection are stored. In fact, the coordinates of these pixels are stored in vectors for those on the x-axis and those on the y-axis. An arithmetic mean of these coordinates is made, and the coordinates of a point representing the focal point of the detected object are determined. Wanting to position the electric seat at another point in the image, we click on the image at that point. The algorithm calculates the distance from the point where the chair has its wheels to the end point, the chosen target point. Speed calculation helps determine the speed of the wheelchair. The greater the distance between the object and the target point, the higher the speed. As it approaches the "target," the distance becomes smaller and the speed decreases in order to position itself accurately. In order to control the displacement direction, a trajectory is virtually traced from the starting point where the device is located at the target point of the calculations (Fig. 7).

The patient being in the target point on a standard wheelchair is in charge of bringing the electric wheelchair at the current point and equipped with add-on mode and navigation system (video camera) through a software application installed on tablet/smartphone.

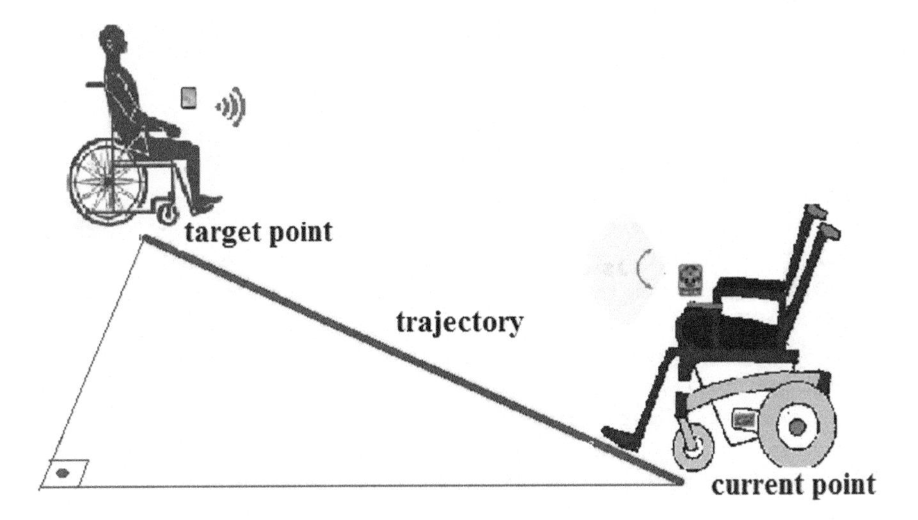

Fig. 7 Explanatory regarding geometric trajectory calculation

The program sequence below shows how to determine the desired trajectory:

```
procedure TForm1.Timer1Timer(Sender:TObject);
var i,semn,j:integer;unghi_dif:TReal;c,d:TCoordonate;
    perioada,viteza_inaintare,rotatie:byte;
beginTimer1.Enabled:=false;GetActualPosition(self);
c.x:=0;c.y:=0;j:=0;//We change the anterior coordinates in the vector
for i:=nrmax downto 1 do
  begincoord[i].x:=coord[i-1].x;coord[i].y:=coord[i-1].y;
c.x:=c.x+coord[i].x*(nrmax+1-i); c.y:=c.y+coord[i].y*(nrmax+1-i);
j:=j+(nrmax+1-i);end;
coord[0].x:=poz_act.x;coord[0].y:=poz_act.y;
//We calculate the arithmetic mean of the anterior positions
c.x:=c.x/j;c.y:=c.y/j;//calculate the variation of the movement to-
wards the anterior positions
d.x:=poz_act.x-c.x;d.y:=poz_act.y-c.y;
viteza:= sqrt(d.x*d.x+d.y*d.y);
if (d.x<>0) then unghi_calculat:=ArcTan(d.y/d.x);
if ((d.x<0) and (d.y>0)) then unghi_calculat:=pi-unghi_calculat;
if ((d.x<0) and (d.y<0)) then unghi_calculat:=pi+unghi_calculat;
if ((d.x>0) and (d.y<0)) then unghi_calculat:=2*pi - unghi_calculat;
unghi_calculat:= round(unghi_calculat*180/pi) mod 359;
d.x:=target.x-poz_act.x; d.y:=target.y-poz_act.y;
distanta:= sqrt(d.x*d.x+d.y*d.y);
if (d.x<>0) then unghi:=ArcTan(d.y/d.x);
if ((d.x<0) and (d.y>0)) then unghi:=pi-unghi;
if ((d.x<0) and (d.y<0)) then unghi:=pi+unghi;
if ((d.x>0) and (d.y<0)) then unghi:=2*pi-unghi;
unghi:=round(unghi*180/pi) mod 359;
unghi_dif:=2*pi-unghi_calculat-unghi;
unghi_dif:=(unghi_dif*180/pi);
unghi_corectie:=(unghi_dif);
//unghi:=unghi+unghi_corectie;
Label1.Caption:=format('%s %.2f', ['Comanda:',unghi]);
Label2.Caption:=format('%s %.2f', ['Calculat:',unghi_calculat]);
Label3.Caption:=format('%s %.2f', ['Corectie:',unghi_corectie]);
Label4.Caption:=format('%s %.2f', ['Diferenta:',unghi_dif])
//ListBox1.Items.Add(format('t:%f p:%f c:%f', [unghi, unghi_calculat,
unghi_corectie]));
viteza_temp:=1+2*distanta*vmax/sqrt(h*h+w*w);
if (distanta < viteza_temp) then viteza_temp:=0;perioada:=100;
viteza_inaintare:=30+round(40*distanta/sqrt(h*h+w*w));
rotatie:=integer(unghi_dif >0)*127 +round(abs(unghi_dif/5));
//if (viteza<30) then rotatie:=0;
if (distanta<((h+w) div 20)) then
```

```
 beginviteza_inaintare:=0;rotatie:=0;end;
//if we am very close to the point
if CPDriver.Connected then
SendData(perioada,viteza_inaintare,rotatie);
PB_F.Position:=viteza_inaintare;
PB_SD.Position:=round(abs(unghi_dif));
if (unghi_dif>0) then L_SD.Caption:=‘>>>>>>>>>‘ else
L_SD.Caption:=‘<<<<<<<<<‘;
```

4 Experimental Results and Discussions

The previous coordinates of the center of the object are also stored, up to 20 points, which are necessary for calculating a trajectory of the object. This trajectory is calculated by performing the arithmetic mean of the coordinates of the previous points; each point having multiplied coordinates with a gradually smaller number, depending on the distance from the current point. This weighting of the coordinates is done so that a point left behind in several positions may not influence the trajectory calculation too much. Depending on the direction of travel, the trajectory of the seat, and the direction in which it should travel, the calculated trajectory, the angle between them is calculated.

If the angle between the two trajectories is greater than 180° then the seat must move to the left, and if the angle is less than 180°, a leftward movement is required. The closer the angle reaches to 180°, the higher the leftward speed is. If the angle is closer to −180°, the speed to the right is higher.

The results of the algorithm that appear on the smartphone’s monitor are illustrated in Fig. 8.

The avoidance of a simple collision is applied in order to stop the mobile wheelchair base when an obstacle is detected. The technical tests on the navigation system have produced satisfying results in terms of person safety and response time.

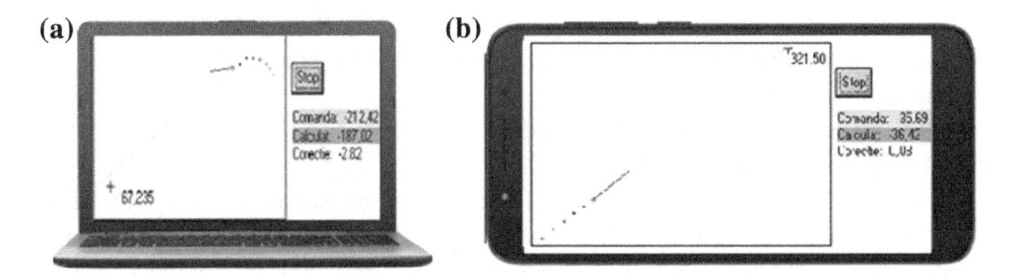

Fig. 8 Explanation regarding the distance and trajectory made with: **a** laptop/tablet having Windows/Linux as operating system; **b** smartphone with Android/IOS as operating system

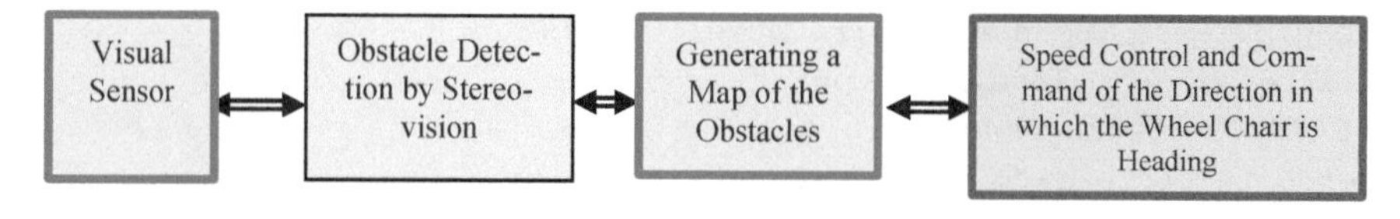

Fig. 9 Functional block diagram of the obstacle avoidance system

The modular solution proposed for the navigation module simplifies the adaptation of the module to various electrically driven wheelchairs.

The experimental results demonstrate the successful integration of all the modules described in the paper. The results were obtained with a prototype running autonomously in simulated environments.

Avoiding obstacles assumes the detection of the obstruction and stopping or changing the direction of the movement for the smart wheelchairs in order to avoid collisions. Sensor systems that could be installed on the smart wheelchair for avoiding obstacles are mainly of the following types: radar, laser scanning, telemeters with ultrasounds and visual sensors.

A diagram of the functional flow for an obstacle avoidance system is shown in Fig. 9.

Data transmitted by each sensor is processed by a detection algorithm and pondered based on sensor performance under a given set of working conditions.

Coordinates of detected targets are integrated into a local obstacle map. The received information is superimposed in the most cohesive manner.

The map is then analyzed to determine the closest obstacle in the direction of the movement. This information is used to slow down the speed of the wheelchair or even to turn it off and to command the motion direction in order to bypass the obstacle.

In making the artificial vision systems the biological vision systems represented source of inspiration.

In this paper we used some ideas and solutions from the field of biological vision. The sight systems mounted on a wheelchair should ensure a high field of vision and an adequate image quality.

Stereo viewing systems play an important role in getting the information about the distance from which various objects in the wheelchair space are located, which can be of great help in avoiding collisions.

Conventional video cameras have a rather limited visual field. For this reason, in the case of wheelchairs it is recommended to use special cameras with panoramic or omnidirectional view.

There are several obstacles that need to be removed so that intelligent wheelchairs may be used on a large scale.

One of the most important technical issues is the cost-versus accuracy ratio that needs to be implemented with the existing market sensors. Until an inexpensive sensor will be designed to detect obstructions and drop-offs in a wide range of operating conditions and surface materials, the preoccupations will limit to smart wheelchairs for the indoor areas.

Another technical problem is the lack of a standard communication protocol between the wheelchair input devices and the wheelchair engine microprocessors.

A standard protocol simplifies the interface between wheelchair technology that is based on intelligent wheelchair development and intelligent wheelchair [26].

Smart wheelchairs are expensive and complicated, so familiarizing and training the passengers will take time and resources. This does not mean that intelligent wheelchair technology cannot be commercialized.

5 Conclusions

Avoiding obstacles is one of the most important issues in creating a smart wheelchair. Without this capability, the wheelchair movement would be restrictive and fragile. The ultimate challenge for the research in this area is the design of a wheelchair that could function autonomously in the real environment, both interior spaces and the outside environment being understood.

The analysis made in the first part illustrates the fact that worldwide the operating wheelchairs are technically outdated. The most economical solution for renewing the wheelchair park is to equip them with smart modules such as the one described above.

The proposed add-on model for attaching to wheelchairs makes life easier for patients, with minimal financial costs and high technical performance. To benefit from versatility and efficiency, the first step is the ability of the electric seat to handle and operate in an unknown environment and to adapt in advance to certain situations. In addition, thanks to the affordability of computer-aided speech recognition technology, this wheelchair generation has the ability to allow voice commands, which is particularly useful in situations where there is a need to assist a patient or in the use of a chair by patients who are not trained.

The navigation system, widely discussed in the paper, is based on panoramic environmental descriptions taken over by a video camera. These environmental images contain the essential elements for the add-on module attached to the rolling electric chair. Because the environment is considered to be of a structured type, the elements can be conveniently described, the wheelchair being equipped with programs that can interpret these concepts and can act accordingly to avoid obstacles, tracking environmental elements, or reorienting according to landmarks. Thus, the functioning of the wheelchair does not depend on a particular environment, it does not need to be predefined, and it simplifies the implementation of the command through a language close to the human one. Such a chair helps and assists people with disabilities, for the manipulation of objects or movements that they cannot perform.

The term "obstacle avoidance" describes a set of software techniques that allow mobile seats to adjust their trajectory and speed according to the environment. Used in conjunction with distance measurement and motion control, the implemented software offers the wheelchairs, reflections similar to living beings and allows them to navigate intelligently.

The ultimate goal is to track the moving chair and move the wheelchair to an arbitrarily chosen point in the image taken from the camcorder after a trajectory that is calculated and automatically generated by the microcontroller. At the same time, necessary corrections are made to the deviations from the calculated trajectory, due to the unevenness of the action area or the intervention of the various external factors. The automated surveillance system, via the microcontroller to which it is closely linked, can keep the data recorded for a long period of time. The communication interface consists of blocks that link the microcontroller and the transmission module of the commands and data reception and respectively transmit-receive modules. The equipment has three main parts: a set of mechanisms (motors, couplers), an interface which connects the drive motors, the command reception and data transmission module, and the receiving/transmitting modules complementing the modules in the communication interface. The video camera depends on the environment and the conditions in which the person with disabilities moves.

Although further research is still needed in order to fully achieve the objectives preliminary tests have shown that the proposed device will offer a more independent life to the patients who equip their electric wheelchairs with this type of modules.

References

1. G. Bourhis, K. Moumen, P. Pino, S. Rohmer, A. Pruski, Assisted navigation for a powered wheelchair, in *Proceedings of the IEEE International Conference on Systems, Man and Cybernetics, Systems Engineering in the Service of Humans*, Le Touquet, France (IEEE, Piscataway, NJ, 1993), pp. 553–558
2. S.P. Levine, D.A. Bell, L.A. Jaros, R.C. Simpson, Y. Koren, J. Borenstein, The NavChair assistive wheelchair navigation system. IEEE Trans. Rehabil. Eng. **7**(4), 443–451 (1999)
3. E. Prassler, J. Scholz, P. Fiorini, A robotic wheelchair for crowded public environments. IEEE Robot. Autom. Mag. **8**(1), 38–45 (2001)
4. R.C. Simpson, E.F. LoPresti, S. Hayashi, I.R. Nourbakhsh, D.P. Miller, The smart wheelchair component system. J. Rehabil. Res. Dev. **41**(3B), 429–442 (2004)
5. M. Mazo, An integral system for assisted mobility. IEEE Robot. Autom. Mag. **8**(1), 46–56 (2001)
6. E. Subhadra, O.F. Andrew, C. Neophytou, L. de Souza, Older adults' use of, and satisfaction with, electric powered indoor/outdoor wheelchairs. Age Ageing **36**(4), 431–435 (2007)
7. R.A.M. Braga, M.P. Beng, L.P. Reis, A.P. Moreira, Intell Wheels: platforma modulara de dezvoltare pentru scaune cu rotile inteligente. J. Rehabil. Res. Dev. **48**(9), 1061–1076 (2011)
8. C.S. Richard, Smart wheelchairs: a literature review. J. Rehabil. Res. Dev. **42**(4), 423–436 (2005)
9. R.C. Simpson, D. Poirot, M.F. Baxter, The Hephaestus smart wheelchair system. IEEE Trans. Neural Syst. Rehabil. Eng. **10**(2), 118–122 (2002)

10. E.S. Boy, C.L. Teo, E. Burdet, Collaborative wheelchair assistant, in *Proceedings of the IEEE/RSJ International Conference on Intelligent Robots and Systems*, Lausanne, Switzerland (Piscataway, NJ, 2002), pp. 1511–1516
11. R. Cooper, T. Corfman, S. Fitzgerald, M. Boninger, D. Spaeth, W. Ammer, J. Arva, Performance assessment of a pushrim activated power assisted wheelchair. IEEE Trans. Control Syst. Technol. **10**(1), 121–126 (2002)
12. D. Cagigas, J. Abascal, Hierarchical path search with partial materialization of costs for a smart wheelchair. J. Intell. Robot. Syst. **39**(4), 409–431 (2004)
13. Y. Matsumoto, T. Ino, T. Ogasawara, Development of intelligent wheelchair system with face-and gaze-based interface, in *Proceedings of the 10th IEEE International Workshop on Robot and Human Interactive Communication*, Bordeaux-Paris, France (Piscataway, 2001), pp. 262–267
14. R.C. Simpson, S.P. Levine, Voice control of a powered wheelchair. IEEE Trans. Neural Syst. Rehabil. Eng. **10**(2), 122–125 (2002)
15. E.M. Craparo, M. Karatas, T.U. Kuhn, Sensor placement in active multistatic sonar networks. Nav. Res. Logist. 287–304 (2017)
16. M.C. Popescu, A. Petrisor, *Robots-Robot Control Systems* (in Romanian) (Ed. Universitaria, Craiova, 2009), p. 139
17. M.C. Popescu, *Telecomunications* (in Romanian) (The Printing House of the University of Craiova, 2005), p. 233
18. A. Lankenau, T. Röfer, A versatile and safe mobility assistant. IEEE Robot. Autom. Mag. **8** (1), 29–37 (2001)
19. M.C. Popescu, *Telecomunications* (in Romanian) (Ed. Universitaria, Craiova, 2008), p. 428
20. T. Gomi, A. Griffith, Developing intelligent wheelchairs for the handicapped, in *Assistive Technology and Artificial Intelligence: Applications in Robotics, User Interfaces and Natural Language Processing*. Lecture Notes in Artificial Intelligence (Springer-Verlag, Heidelberg, 1998), pp. 150–178
21. I. Moon, M. Lee, J. Ryu, M. Mun, Intelligent robotic wheelchair with EMG, gesture-, and voice-based interfaces, in *IEEE/RSJ International Conference on Intelligent Robots and Systems*, Las Vegas, NV (Piscataway, NJ, 2003), pp. 3453–3458
22. J.D. Yoder, E.T. Baumgartner, S.B. Skaar, Initial results in the development of a guidance system for a powered wheelchair. IEEE Trans. Rehabil. Eng. **4**(3), 143–151 (1996)
23. M.C. Popescu, A. Petrisor, 2D tracking control algorithms, in *Proceedings of the 5th WSEAS International Conference on Computational Intelligence, Man-Machine Systems and Cybernetics*, Venice, Italy, pp. 368–373 (2006)
24. A. Petrisor, N.G. Bizdoaca, M. Drighiciu, M.C. Popescu, Control strategy of a 3-DOF walking robot, in *The International Conference on "Computer as a Tool"* (IEEE, Warsaw, 2007), pp. 2337–2342
25. H.N. Chow, Y. Xu, S.K. Tso, Learning human navigational skill for smart wheelchair, in *IEEE/RSJ International Conference on Intelligent Robots and Systems*, Lausanne, Switzerland (Piscataway, NJ, 2002), pp. 996–1001
26. X. Li, X. Zhao, T. Tan, A behavior-based architecture for the control of an intelligent powered wheelchair, in *IEEE International Workshop on Robot and Human Interactive Communication*, Osaka, Japan (Piscataway, NJ, 2000), pp. 80–83
27. A. Petrisor, N. Bizdoacă, A. Drighiciu, M.C. Popescu, Three legs robot—application for modelling and simulation of walking robots control algorithms, in *The 3rd International Conference on Robotics*, Buletinul Institutului Politehnic din Iasi, Editat de Universitatea Tehnica "Gh. Asachi", Tomul LII (LVI), Fascicula 7B, Sectia Constructii de masini, pp. 127–133 (2006)
28. S.P. Parikh, R.S. Rao, S.H. Jung, V. Kumar, J.P. Ostrowski, C.J. Taylor, Human robot interaction and usability studies for a smart wheelchair, in *IEEE/RSJ International Conference on Intelligent Robots and Systems*, Las Vegas, NV (Piscataway, NJ, 2003), pp. 3206–3211
29. H. Wakaumi, K. Nakamura, T. Matsumura, Development of an automated wheelchair guided by a magnetic ferrite marker lane. J. Rehabil. Res. Dev. **29**(1), 27–34 (1992)

30. H.P. Moravec, Certainty grids for mobile robots, in *NASA/JPL Space Telerobotics Workshop* (Pasadena, CA, JPL Publications, 1987), pp. 307–312
31. S.B. Shuvra, *Handbook of Signal Processing Systems* (Springer International Publishing AG, 2019)
32. M. Sun, J. Hu, An image edge feature extraction method based on multi-operator fusion. Rev. Téc. Ing. Univ. Zulia **39**(10), 331–339 (2016)
33. J. Lee, H. Tang, J. Park, Energy efficient canny edge detector for advanced mobile vision applications. IEEE Trans. Circuits Syst. Video Technol. **28**(4), 1037–1046 (2018)
34. C. Poynton, *Digital Video and HD*, 2nd edn. (Morgan Kaufmann, Burlington, 2012), p. 134
35. Clean PNG, https://www.kisspng.com/png-rgb-color-model-rgb-color-space-cube-white-colour-724188/download-png.html. Accessed June 2019
36. Y. Zheng, Y. Chang, M. Sarem, Accurate computation of geometric moments using non-symmetry and anti-packing model for color images. Int. J. Comput. Commun. Eng. **6**(1), 19–28 (2017)
37. C. Cercel, *Command the Position of an Object in a Plane Through Image Analysis* (in Romanian), pp. 35–52 (2006)
38. R. Braniscan, M.C. Popescu, A. Naaji, Secure PHP OpenSSL crypto online tool. Int. J. Adv. Comput. Netw. ITS Secur. **5**(2), 108–112 (2015)
39. V. Chernov, J. Alander, V. Bochko, Integer-based accurate conversion between RGB and HSV color spaces. Comput. Electr. Eng. **46**, 328–337 (2015)

An Innovative Prediction Technique to Detect Pedestrian Crossing Using ARELM Technique

A. Sumi and T. Santha

Abstract Monitoring Systems of Automobile Industries and Surveillance Systems use operations based on computer vision to identify objects in motion. Most such applications that employ, pattern identification techniques to detect person on road are done through feature mining and classifier development framework. A learned classifier is arranged over the method of recognizing features that are extracted from the video frames. In this paper, classification of pedestrian features is performed and subsequently the presence of pedestrians is predicted. A new classifier, named Asymmetric Least Squared Approximated Rigid Regression Extreme Machine Learning [ARELM] is proposed for the classification and prediction purposes. This classifier combines the strengths of aLs-SVM that deploys the expectile distance as the measurement for boundary values and RELM in handling the multi collinear data. The proposed classifier improves the accuracy in detecting the pedestrians among the navigating things and ensures better prediction, on comparisons with existing classifiers like SVM, BPN used for the same applications.

Keywords Pedestrian detection · Machine learning · ELM · ARELM · Asymmetric Least Squares

1 Introduction

Progress in the development of intelligent automobiles and autonomous vehicles without the intervention of human beings has led to the significance of smart assistances to humans in recent days. Consequently, the visual experiences of human beings face difficulties in identifying the pedestrian in diverse environment [1]. Identification of person on road, a recognized subdivision, of computer vision has been in research extensively for the past two decades. The mentioned process

A. Sumi (✉) · T. Santha
Department of Computer Science, Dr. G.R. Damodaran College of Science, Coimbatore, Tamil Nadu 641014, India
e-mail: sumianandhphd@gmail.com

D. J. Hemanth (ed.), *Human Behaviour Analysis Using Intelligent Systems*, Learning and Analytics in Intelligent Systems 6,
https://doi.org/10.1007/978-3-030-35139-7_6

can be used in numerous areas like monitoring with the video using CCTV and autonomous automobiles. Detection method is known to be the competent technique for particular object identification, since pattern identification is broadly implemented in varied fields such as face recognition, pedestrian identification, automobile identification as well as article finding and categorization.

ITS (Intelligent Transport System) has fascinated many researchers for the past few years. The term ITS, refers to the framework with the sensors installed over the automobile for processing and assessing the protection provided to the pedestrian in various circumstances on the road. Numerous investigators constructed and recommended street path identification technique [2], automobile identification strategies [3] and traffic symbol realization procedure [4] with the help of numerous image computing strategies incorporated in ITS. Specific strategies to identify the persons on road in an involuntary manner [5] must be imperative for safeguarding the human beings inside the vehicles and those walking on the road from involving in accidents.

Identifying the persons on the road will involve a set of complex issues since the existence of diverse styles of humans, the color of their clothes, movement behavior on the road, alteration in the lighting effect in the circumstances, etc. [6]. Even though the identification of persons over road attained exponential enhancement for the past few years, complicated blockages that will occur in real-time is the impediments faced by the system. As per investigations carried out, around 70% of the persons on roads will be blocked by the objects for visual experience at least in a single frame in the videos [7].

Xu et al. [8] introduced the notation of pedestrian crossing event which is shown in Fig. 1. This explains the sudden crossing of pedestrian in front of vehicle by spatio-temporal volume is the ratio of visibility of pedestrian in the camera and given by the notation α and is define as

$$\alpha = \frac{X_e - X_r}{W}, \tag{1}$$

where, X_e is the X axis value of the right edge of the pedestrian's bounding box. X_r is the X axis value of a vertical reference line, and W is the horizontal width of the

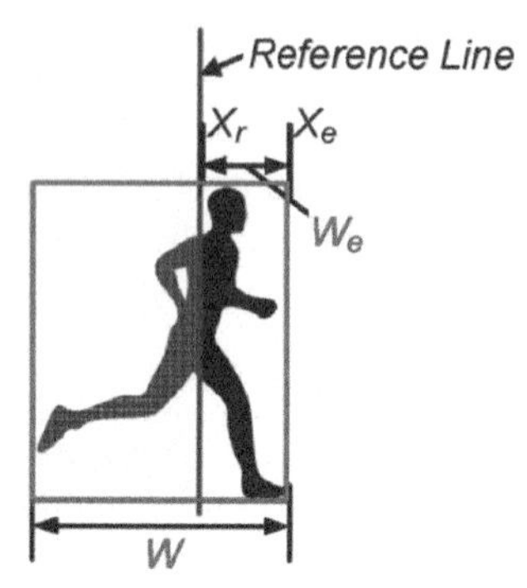

Fig. 1 Notation for a pedestrian crossing event [8]

bounding box. In the entering ratio equation "entering width" is denoted as $W_e = X_e - X_r$. By using this ratio of pedestrian entering into the visibility of camera the spatio-temporal cubic of a pedestrian crossing event starts from a pre-defined threshold α_e and ends when the entering ratio reaches a certain threshold α_l.

Numerous classification strategies, features and distortion prototypes were utilized for the purpose of accomplishing the development of identifying human beings on roads. Nevertheless, evolving outcomes pertaining to the continued researches were recommended in constructing and employing extremely consistent and strong human detection capacities, endure to be tremendously thought-provoking. Commercially accessible human detection strategies available at present and other frameworks expected to be fruitful to a maximum level in organized circumstances in which corporeal and ecological features will be recognized and regulated [9].

Clear, quick identification of objects will be significant for the triumph of upcoming evolution of automatic visual frameworks. The aim of this principle is to concentrate on the development of identification, which might be effortlessly manageable with any fresh arena or circumstance. An all-purpose human identification framework, incorporate the feature of making the system learns to identify diverse categories of humans. So it is obligatory to concentrate on the development of pedestrian detection system that can be trained under various circumstances. This framework learns with an illustration that provides permission to circumvent the requirement for the physically developed solution. The fundamental strategy of identifying the person on road, not depend on the knowledge of navigation and the circumstance of the picture or the number of substances existing in the scene [10].

Humans are detected from the videos are considered a significant proportion of numerous circumstances. Majority of conservative strategies have relied on the use of physical characteristics that will be based on the nature of the issue and suitability for particular situations. Besides, the above-mentioned features are exceedingly vulnerable to emerging activities like alterations in lighting, camera faults, and discrepancies caused by the dimensions of the substances. The proposed feature training strategy is cost-effective and effortless since exceedingly intellectual and distinguishable characteristics are generated automatically without the need of professional skills. In the current work the use of automated learning methods for human identification is accompanied by videos made with dynamic camera with variable features [11].

There are two primary kinds of feature learning method that are supervised and unsupervised learning. A supervised learner needs input/output pairs to learn the features. An unsupervised learner uses only inputs to find its features. Accuracy and speed of learning are the important parameter to be considered for obtaining the "person", "non-person" categorization. In the past few years most of the researchers have focused on image description for pedestrian detection. Extraction of features technique is segregated into two groups, based on handmade and training with the help of the computer. Histograms of Oriented Gradients are the most intriguing extraction by manual crafting [12]. HOG shows the local gradient magnitude and direction of the picture that normalizes vector characteristics blocks, based on the gradient characteristics. In addition, HOG allows communication between blocks that respond to changes produced in lighting with the least amount of alteration. This might competently demonstrate the edge feature of an individual's body. Generally utilized techniques that fall under the category of manual handcrafting techniques are listed as follows: 1. Haar-like [13], 2. SIFT [14], 3. Covariance descriptors [15], 4. Integral channel features and 5. 3D geometric characteristic [16].

With advancement in the processing capacity of computers and a enormous increase in the amount of information, the development of computerized involuntary learning methods by multiple researchers has gradually increased. Sermanet et al. [17] projected the ConvNet framework that utilizes the real values of the pixel from the image will be considered as the input. The work incorporates both unsupervised and supervised techniques in training of the multi-phase involuntary computerized scarce convolution encoder. This technique produces relatively notable results through the operation of the INRIA pedestrian database; on the other hand, it is difficult to achieve high quality results with Caltech Dataset. UDN [18] termed as the combined deep neural network framework incorporated alongside distortion in addition with obstruction prototype, that accomplishes the least rate of non-identification of the target when compared with conservative manual identifiers for recognizing the individuals HOGCSS SVM technique on Caltech and ETH dataset. Lim et al. [19] utilize the supervised learning framework for the extraction of features with medium quality depending on the knowledge of contour and learn with random forest classifier for improving the functionality of the projected technique.

The extraction of feature provides a better description in the identification of individuals walking on the highway; on the other side, manual feature extraction techniques are not able to train with significant characteristics and contain the least flexibility. Computerized methods are capable of extracting the features of the pedestrian by utilizing techniques such as feedback propagation. These require a lot of learning illustrations and consume time for training. It requires a greater bounding of hardware [20].

Convolution Neural Network accomplished the huge quantity of competence in identification of the persons walking on the roads. CNN reliant techniques are divided into two significant computing steps for the proposed features in addition with classification technique. Initially, conservative methods in the detection of individuals walking on the road were used to obtain the suggested applicants. Subsequently, the suggestions will be classified as "pedestrian" or "non-pedestrian" with the help of CNN. Regardless of the achievement of huge competence, CNN has to be enhanced considerably as: (1) Most of the techniques utilize the final phase features of CNN by using SVM to classify the suggestions. However, the distinct phases of CNN signify the diverse features of image characteristics. Initial phase describes the local fluctuation of the image, while in the final phase it utilized the global information for extraction. Every phase with CNN comprises of diverse differentiable features that is utilized in training the classifier. (2) Certain techniques will utilize conservative approaches which rely on the physical extraction of features in creating the recommendation of the candidates whereas they avoid the recommended scores. (3) Deep CNN will consume more time for the processing due to complex calculations reliant on convolution operations.

Recently, researchers have carried out some research to find solutions to the above-mentioned issue. Li et al. [21] suggested training, which connected many CNN frameworks with diverse purposes. The Least-resolution CNN, discards several background regions at the primitive stage, circumvents screening of the complete image with greater resolution. As a result, CNN subsequently minimizes the expense incurred for the processing. Nevertheless, the learning procedure of many of the CNN frameworks will be comparatively complicated procedure [22].

Most of the HOG and the improved version of HOG are developed and utilized for ADAS. An improved and customized HOG descriptor was given by Ameur et al. [23] with simple linear SVM to implement ADAS easily. This work focused on precise identification of humans from the video frame under dynamic conditions and enhances the speed of computation that pertains to the technique of identification of the persons on road.

2 Literature Survey

Quick and reliable video-dependent identification of street users will be the dynamic investigation field that belongs to automatic "visual years" for the last two decades. At present, it was measured as the dynamic field of research and also the most complicated one. Irrespective of the existence of a number of state-of- the-art techniques which operate significantly over benchmark datasets, the objective of achieving satisfactory functionality in the identification of pedestrians will be much-needed. Previous researches in this area established the kernel Support Vector Machine-dependent classifiers might be combined the least challenging integer. This feature of acquiring authoritative identifiers that achieve accuracy identical to the current cascade-dependent identifiers that execute complex features requiring floating-point features. Conservative techniques involved in the identification of pedestrians necessitate skill by the individuals to explain the characteristics of pedestrian and, at last, to incorporate them alongside the classifiers. Currently, the employed technology of deep learning, particularly Convolutional Neural Networks (CNN) over the image and audio processing revolutionized real-time processing and achieved greater competence. This evokes a significant concentration of researchers on the component of deep learning. Artificially constructed approaches for the extraction of characteristics based on conservative processes will involve a faulty explanation of the situation of the pedestrian, which is extremely complex in terms of the context. Research by Li et al. [24] established the strategy for the identification of pedestrians that is completely reliant on Deep Convolutional Neural Network alongside many layers. The proposed strategy exploits the complete benefits of DCNN and takes the features from the database that is useful in the identification of pedestrian. During the area projection phase, they utilized the edge blocks instead of sliding window techniques to take the windows for further processing. By performing this operation, the problem of generating redundant windows is overcome by conservative techniques. Finally, they have acquired few windows of the highest quality, which are deemed to have a greater importance for the purposes of the classification activity that follows. They performed multi-group analysis and investigation under the proposed framework. This analysis demonstrates that identification framework of pedestrian depends on deep learning performed in a superior way in compared to some of the conservative techniques on features that are handcrafted and learned automatically.

Numerous research utilized in the recognizing background versus foreground errors, that profoundly impacts the quality of identification. In the paper by He et al. [25] prominent involvement in formulating the areas of interest will be the one that includes semantic information interest that grossly minimizes background versus foreground mistakes that are identified by the foreground target. Initially, the complete Convolutional Neural Network will be used to create a pedestrian heat map from the input image, which is learned with the help of on the Caltech Pedestrian Dataset. Subsequent utilization of process that includes morphological computation of the images, semantic area of attention will be taken out from the heat map. Eventually, complete images are segregated into foreground and background with the help of information about the semantic areas of interest in order to support the identification process of the pedestrians. They evaluated their strategy with the help of above-mentioned dataset. Identifiers with changing proportions of enhancement are achieved with the help of the desired area that contains semantic information. The Greatest will be the one that surpasses the advanced one. The work enhances the HOG + SVM pedestrian identifier in the desired areas with semantic information which is achieved with the help of the completely-convolution neural network. Learning end-to-end and pixel-to-pixel, complete convolution networks might differentiate the background and foreground competently in minimizing the errors between the background versus foreground errors of the detectors. Concurrently, the rapidity of the technique will be enhanced successfully. In addition, they analyzed the desired areas with semantic information along with some other identifiers like ACF, Joint Deep, Spatial Pooling and LDCF++, by using the Caltech Pedestrian Detection Benchmark dataset. Depending on the investigations that are performed in the required fields, the accuracy of the semantic data is enhanced.

A similar research by Bilal and Hanif [26] established the unique strategy for employing the soft cascade for the purpose of accelerating the processing by the kernel classifier. The projected strategy accomplished fast decline of the actual negatives at the primitive phase itself via detecting the appropriate characteristic elements organized with the help of predicted energies of the conforming kernel operations. The above-mentioned strategy incorporated, along with the addition of manifold identifiers, with diverse measures and hardware assistance for concurrent operation and vector calculation. These lead to the three folds of the computing speed related to the currently available identifiers and also accomplish the improvement in accurate proportional identification.

Identification of persons on road developed has become the most significant issue because of the involvement of the techniques in several areas of applications like surveillance, driverless automobiles, framework tracking, and robotics. Consequently, Nguyen et al. [27] established a competent technique that identified the humans on road with the help of Support Vector Machine (SVM) and the Histogram of Oriented Gradients (HOG) characteristics. The work established the framework which had the capacity to minimize the group of member identification areas and categorizing pedestrians and non-pedestrians in city traffic circumstances with the help of the stixel world and HOG + SVM, correspondingly. In the suggested technique, the stixel world calculation will be used to determine the desired input region. In relation to the approach used to categorize the dataset, they have been effective in creating a new dataset that encompasses around images of 3100 people.

Supreeth and Patil [28] suggested an innovative technique to pursue object using flexible linear Support Vector Machine (SVM) and RGB Mean filter. Histogram of Oriented Gradient (HOG) characteristics was taken out for identifying the substance in the video frames. The substance and the circumstantial HOG characteristics are utilized for making the SVM to learn. The learned SVM classifier identifies the substance in consecutive frames and improves the route of the substance. The SVM informs the characteristics of context after passing ever 20 frames. The SVM evaluation rank is huge while the precise substance was determined. In addition to the SVM classifier, the RGB Mean Filter was used while the assessment rank was developing the least value. This was examined with numerous videos could pursue substances competently. The suggested technique was analyzed using video capture on camera navigation, substance structure distortion, blockings and fluctuating in lighting conditions for testing the strength.

Regardless of current substantial development in the area of identification of pedestrians remain to be a tremendously thought-provoking challenge in actual circumstances. Recently, certain researches demonstrated the benefits in utilizing amalgamations of portion/patch-dependent identifiers in order to tackle the issue of huge inconsistency of postures and the presence of incomplete blockages. From the beginning of 2017, deep forest technique was proposed for overcoming the issue of blanking of the decision tree in the area of deep learning. Deep forests had the least characteristics when compared with deep neural network and the benefits that pertain to greater classification accuracy. Zheng et al. [29] suggested the unique identification of pedestrian, technique which incorporates the adaptability of the portion-dependent prototype along with the faster processing duration of the deep forest classifier. In the suggested grouping, the portion of the part assessments will be considered with the help of local assessment skills at the nodes of the decision tree. Initially it performed by choosing the features that depend on Extreme Learning Machines for obtaining feature groups. Subsequently, they utilized the deep forest to categorize the feature groups for obtaining the rank, which is the result of the local qualified persons.

The work by Navarro et al. [30] established the computerized sensor-dependent framework for identifying the pedestrians with the driverless automobile area. Even though the automobile was furnished with a wide range of sensors, the paper concentrated on computing the statistics provided by the Velodyne HDL-64E LIDAR sensor. The cloud of themes produced with the help of the sensor was computed for identifying the persons on road by means of choosing cubic structures and enforcing automatic visualization and machine learning techniques to the XY, XZ, and YZ projections of the points encompassed in the cube. The work compares the comprehensive investigation with the functionality of three diverse machine learning techniques: k-Nearest Neighbors (kNN), Naïve Bayes classifier (NBC), and Support Vector Machine (SVM). The above-mentioned techniques were learned with 1931 samples. The concluding functionality of the technique considered the actual traffic circumstances that encompassed 16 pedestrians and 469 samples of non-pedestrians, demonstrates sensitivity (81.2%), accuracy (96.2%) and specificity (96.8%).

The work by Ouyang et al. [31] established the integrated deep structure which gets trained together with four elements such as feature mining, the management of distortions of blockages and categorization—for the purpose of identifying the persons on road. Through communication between the above-mentioned codependent elements, the combined training carried out an improvement in the accuracy of identification over the standard datasets, which were very exclusively used for the identification of persons on the road. Comprehensive investigation assessments evidently demonstrate that the suggested innovative framework could boost the robustness of every element while entire elements collaborate with one another. They empowered the deep framework with the help of establishing the distortion layer, which contains the improved adaptability for integrating several distortions dealing strategies.

Several strategies, depending on the mining of characteristics and classification have been developed for the past few years. The work by Errami and Rziza [32] established the innovative strategy for the identification of persons on road depending on supervised classification. They suggested the utilization of fundamental probabilistic functions for adjusting the Support Vector Regression (SVR) for the purpose of binary classification. The classification chain followed in the paper was described as trails: Initially, they utilized Haar wavelet decomposition and Histograms of Oriented Gradients (HOG) for the purpose of mining the characteristics. For the purpose of classification, they utilized the proposed technique of SVR and analyzed with KNN and SVM classifiers. Research has been carried out with the government pedestrian information set. The results achieved proved the enhanced functionality of the proposed classification approach.

An innovative object pursuing technique was suggested by Wang et al. [33] which exploited the benefit of the quick learning capacity of Extreme Learning Machine (ELM). In particular, the following object was perceived as a binary classification issue and ELM was used to determine the optimal isolated hyper plane between the substance and the circumstances competently. For accomplishing improved strength, following two conditions were established in ELM training: (i) target pictorial fluctuations that existed in various frames, were smooth (i.e. smoothness) and (ii) likelihood for actual object of image samples about the pursued target route was desired in comparison with the circumstantial ones (i.e. preference). Research with stimulating arrangements has determined that the intended follower has achieved constructive results when compared to sophisticated techniques.

"Target Pursuit" will be one of the most challenging operations in the field of automatic visualization frameworks. The mentioned technique concentrates on the identification and following the specific objects in arrangements. Fluctuations in lighting circumstances, navigation of the target, blockages and circumstantial confusion have made tracking the target tremendously complicated. Yu et al. [34] proposed an innovative technique that depends on the Extreme Learning Machine (ELM). The mentioned approach contains the following divisions: learning, pursuing and improvement of the classifier. The leaning phase focus on learning the ELM with the use of training set. HOG features are extracted in the initial frame of every arrangement to make the ELM learn. Subsequently, the following phase develops an estimation of the object location and identifies the target in member areas. An easier object navigation prototype was formulated to predict the location of the object. In conclusion, by conferring the following results, the classifier could be enhanced for operational practice.

The work by Kharjul et al. [35] established the strategy for the identification of pedestrian's framework to minimizing the pedestrian—vehicles accident by providing safety vehicle system. This is significant for identifying the persons on road, competently and correctly in numerous automated visualization areas, such as smart conveyance structures and secured support provided to humans who are involved in negotiating the automobile successfully. The work establishes the identification of persons depending on images. They utilized AdaBoost technique and flowing techniques for segregating the pedestrian members from image. Support vector machine (SVM) is utilized and trained by taking the gray images and edge images features as input.

Yang et al. [36] established the strategy for identification of persons on road with the thermal infrared image depending on Binarized Normed Gradients (BING) and ELM. The members were chosen with the help of BING and the characteristics of Histogram Oriented Gradients (HOG) and pedestrian appearance were utilization in ELM for identifying the position of the persons on road in thermal infrared image. The investigational outcomes demonstrate that the projected technique outclasses the conservative SVM approach. Nevertheless, the occurrence of certain issues was observed. Foremost, the mentioned technique could not differentiate the persons on road who were nearer to each other, subsequently certain members, e.g. trees, cars, and street lamps might occasionally appear as pedestrians. They might continuously optimize the mentioned technique to determine the solutions to the issues the problems.

3 Problem Statement

Alongside the steady intensification pertaining to traffic in the road, the probable risk of accidents also upsurges too. Annually around 1.2 million individuals of public face the risk of accidents and a considerable number of them are victimized to fatal effects [1]. Statistical data about road accidents throughout the year showed that about 10 million individuals were involved in the accident. Consequently, identification of persons on road will be the issue that takes a substantial amount of communal, commercial and systematic concentration. Nevertheless, the mission of identification of persons on road demonstrates to be complicated because of numerous conditions, such as the attainment of circumstances, interior and exterior disturbances, pedestrian inconsistency according to dimension, and category of dressing and location of the image.

Subsequent to wide-ranging and profound examination of preceding analysis of reviews constructed in the background, the classification-dependent strategy appears to be the majority of the methods that are being utilized. It comprises two important phases: features mining and supervised training (classification). Every image in the database about pedestrians might be explained utilizing the feature extractor (descriptor), and subsequently the vector mined might be provided as the input to the classifier for learning.

4 Proposed Methodology

Considered to be the major significant challenges, the provision of safety to road users, who are more susceptible to accidents and the reasonable entry space for the conveyance framework, are encountered throughout the world recently. Therefore, the automatic detection of pedestrian is the most essential safety measure for all road users.

Figure 2 shows the block diagram of proposed methodology. The proposed methodology contains three significant steps. Firstly, the conversion of video image into a frame and the application of appropriate preprocessing technique to the acquired image are performed to improve image display by means of image enhancement. Secondly, the most significant step that involves feature extraction technique. This extracts the features connected with humans on roads in the presence of occlusions, poor lighting conditions, distortions happened either because of the non-linearities of sensor or because of the circumstance at which the image is acquired. Finally, the third step includes the prediction of presence of pedestrians with the help of machine learning classifier. This is achieved by training the system with extracted features.

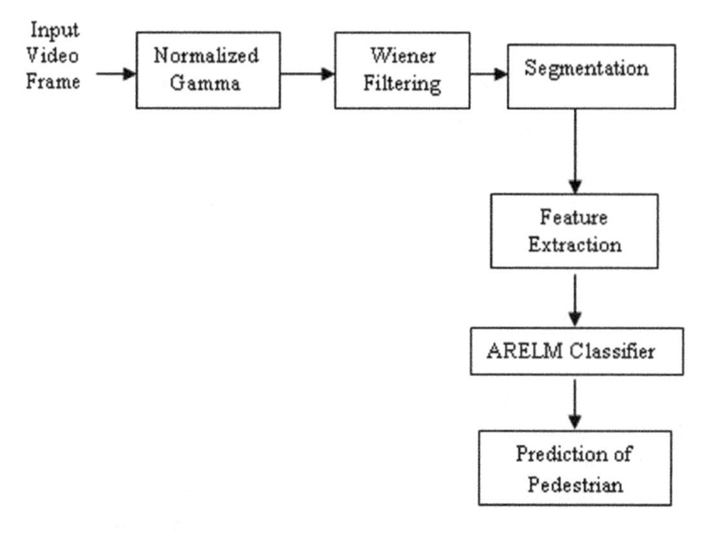

Fig. 2 Block diagram of proposed methodology

4.1 Image Preprocessing

The captured video gets affected by motion blur due to the movement of sensors and objects. In the preprocessing stage, the motion blur is removed using the Wiener filter and further evaluated by histogram equalization.

Enhancing the visualization of video display is one among the significant issue encountered while evaluating the frames in video. For the purpose of recognition, improving the information from the video is most significant, in order to obtain the absolute visual display. Contrast improvement in self-augmentation will be utilized in computing the local contrast in diverse areas that belong to the images provided. Histogram equalization will be utilized for the purpose of improving the contrast [37].

In pedestrian recognition, distortion due to navigation happens while taking the picture using motion camera. Therefore the filter should be utilized for eliminating the distortion [38]. Wiener filter is employed since it would help in the noise reduction and provide greater compensation during image retrieval. The distortion procedure y (m, n) is developed through appending white navigation noise along with the occasioned input video computed by h (m, n) filter is shown in Fig. 3.

4.2 Feature Extraction

Collection and mining of features will lead to the precise recognition of the public on road by circumventing the false recognition proportionately. For this a new

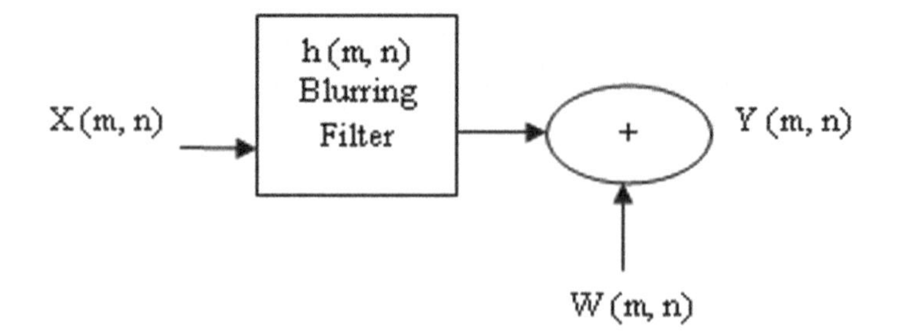

Fig. 3 Removal of motion blur

frame level features are utilized. This feature will be taken into consideration along with the features in the frame of video, which produce the abridged form of video-level description.

Focusing on the choice and taking out appropriate features are performed by means of preserving the variations which occur in two consecutive frames and the variation will be reserved as the feature. The sequential knowledge detained by means of dispensed frame-level features with feature translation strategy is adopted in some training approaches. The alteration between the consecutive frame sets are taken as input for the framework and this contains the capacity to accomplish the inferences noticeably reliant on the inclination of varied features in neighboring frame [39].

4.3 Classification

A classifier based on machine learning is used to predict the presence of pedestrian by taking the frame level features extracted from enhanced video. In order to meet out increased accuracy rate and high detection speed, Asymmetric Least Squares RELM classifier is employed in this paper.

Asymmetric Least squares RELM
The proposed classifier utilizes the benefits of Asymmetric Least Square SVM that takes into account the expectile distance as the dimension of boundary rather than exploiting the least value and the compensations of RELM in dealing with the multi collinear data.

Basic ELM technique [40] can be summarized as follows.

Step 1. Allocate random input weights W and bias of concealed layer nodes B.
Step 2. Compute the hidden layer output matrix H.
Step 3. Compute the output weights: $\hat{\beta} = (H^T H)^{-1} H^T T$.

The above-mentioned steps are involved in the concept of normal ELM technique. In this connection, certain characteristics of the solution must be explained. While procuring the data, the step (3) has to be adjusted:

$$T = H\beta + \varepsilon \tag{2}$$

where $\varepsilon \in (0, \sigma^2)$ characterizes the prototype ambiguity or noise distortion. The improved solution is

$$\hat{\beta} = \left(H^T H\right)^{-1} H^T T = \frac{\sum_{i=1}^{n} H_i^T (H_i \beta_i + \varepsilon_i)}{\sum_{i=1}^{\hat{N}} H_i^2} = \beta + \frac{\sum_{i=1}^{\hat{N}} H_i^T \varepsilon_i}{\sum_{i=1}^{\hat{n}} H_i^2} \tag{3}$$

1. $E\left(\hat{\beta}\right) = \beta$
2. $V(\hat{\beta}) = \frac{\sigma^2}{\sum_{i=1}^{\hat{N}} H_i^2} = \sigma^2 \sum_{i=1}^{\hat{N}} \frac{1}{\lambda_i}$ where λi is the ith eigenvalue of $H^T H$
3. $MSE\left(\hat{\beta}\right) = \frac{1}{N} E\left(\hat{\beta} - \beta\right)^T \left(\hat{\beta} - \beta\right) = \frac{1}{N}\left(\frac{\sigma^2}{\sum_{i=1}^{\hat{N}} H_i^2}\right) = \frac{1}{N}\sigma^2 \sum_{i=1}^{\hat{N}} \frac{1}{\lambda_i}.$

Assume $H^T H$ might not be nonsingular; that is to deliberate while the H matrix is multicollinear, certain eigenvalues might converge to zero, whereas $(\hat{\beta})$ and MSE $(\hat{\beta})$ might develop greater, and it might impact its constancy and simplification. Usually, the data obtained from the arena are inclined to contain the existence of multicollinear issues. In the subsequent section, one of the foremost outcomes shall be established. Similar to the concept of Ridge Regression to tackle the multicollinear issue in least square strategy, it is termed as "enhanced technique R-ELM".

Improved ARELM Algorithm

As deliberated above, the ELM technique faces great unpredictability and worst functionality when tackling the multicollinear data. In the projected R-ELM technique, LDL^T decomposition will be enforced against the symmetric matrix $H^T H$. LDL^T decomposition (also called LDL^T factorization) factors the symmetric matrix as the multiplication of the least triangular matrix (L), a diagonal matrix (D), and the transpose matrix of the initial least triangular matrix (LT). While applying the decomposition, it will be framed with the least threshold to certain singular elements of D matrix that overcomes the issue of the puzzle of multicollinear data.

The LDL^T decomposition of a symmetric matrix ($H^T H$) will be as trails:

$$H^T H = LDL^T \tag{4}$$

While

$$L = \begin{matrix} 1 & 0 & \dots & 0 \\ 1_{21} & 1 & \dots & 0 \\ . & . & & . \\ l_{\hat{N}1} & l_{\hat{N}2} & \cdot & 1 \end{matrix}$$

$$D = \begin{matrix} d_1 & \dots & \\ \vdots & \dots & \vdots \\ & \dots & d_{\hat{N}} \end{matrix}$$

Every component of L and D will be successively intended utilizing a repetitive strategy; that will be

$$d_k = a_{kk} - \sum_{m=1}^{k-1} u_{km} l_{km} \tag{5}$$

$$u_{jk} = a_{jk} - \sum_{m=1}^{k-1} u_{jm} l_{jm} \qquad j = k+1, k+2, \dots \hat{N} \tag{6}$$

$$l_{jk} = \frac{u_{jk}}{d_k} \tag{7}$$

where a_{jk} is the elements of the ordinary matrix $H^T H$.

Subsequent to LDL^T decomposition, if the actual matrix has multicollinear issue it will be found out with the help of the values of matrix D. If values of certain components are nearer to zero, the ordinal matrix will be multicollinear. So as to acquire strong matrix, an improved technique for computing the values of L and D will be provided as trails:

$$d_k = \left\{ \begin{matrix} a_{kk} - \sum_{m=1}^{k-1} u_{km} l_{km} \\ sgn\left[a_{kk} - \sum_{m=1}^{k-1} u_{km} l_{km}, \varepsilon_0 \right] \end{matrix} \left| \begin{matrix} if \left| a_{kk} - \sum_{m=1}^{k-1} u_{km} l_{km} \right| > \varepsilon_0 \\ else \end{matrix} \right. \right\} \tag{8}$$

$$u_{jk} = a_{jk} - \sum_{m=1}^{k-1} u_{jm} l_{jm} \quad j = k+1, k+2, \dots \hat{N}$$

$$l_{jk} = \frac{u_{jk}}{d_k} \quad k = 1, 2, \dots \hat{N}$$

while ε_0 is a suitable positive number.

Subsequent to decomposition, the output weights might be computed in the succeeding innovative manner:

$$\hat{\beta} = \left(LDL^T\right)^{-1} H^T H = \lambda_0 \beta + \frac{\sum_{i=1}^{\hat{N}} H_i^T \varepsilon_i}{\sum_{i=1}^{\hat{N}} L_i D_i L_i^T} \tag{9}$$

while $\lambda_0 = \frac{\sum_{i=1}^{\hat{N}} H_i^T H_i}{\sum_{i=1}^{\hat{N}} L_i D_i L_i^T}$ whose value will be nearer to 1.

Assume that the matrix D can be decomposed into two divisions. That will be, $D = D^{1/2} \cdot (D^{1/2})^T$, and if certain components have negative values, complicated decomposition has to be taken into account. Subsequent recomputed functionality markers are:

1. $E\left(\hat{\beta}\right) = \lambda_o \beta$
2. $V\left(\hat{\beta}\right) = \frac{\lambda_o \sigma^2}{\sum_{i=1}^{\hat{N}} (L_i D_i^{\frac{1}{2}})(L_i D_i^{\frac{1}{2}})^T} = \lambda_0 \sigma^2 \sum_{i=1}^{\hat{N}} \frac{1}{\lambda_i}$
   while λ_i is the ith eigenvalue of (LDL^{T}).
3. $MSE\left(\hat{\beta}\right) = \frac{1}{\hat{N}} E\left(\hat{\beta} - \beta\right)^T \left(\hat{\beta} - \beta\right) = \frac{1}{\hat{N}} (\lambda_o - 1)\beta^2 = \lambda_0 \sigma^2 \sum_{i=1}^{\hat{N}} \frac{1}{\lambda_i}.$

Framing the threshold to the values of each component in matrix D can develop confirmation that $|\left(\hat{\beta}\right)| \leq \hat{N}\lambda_0\sigma^2/\varepsilon_0$ and $|\text{MSE} \left(\hat{\beta}\right)| \leq 1/\hat{N}[|(\lambda_0 - 1)\ \beta^2| + \lambda_0\sigma^2/\varepsilon_0]$. Additionally, the approximation of $\hat{\beta}$ is not unbiased as recompence $(\lambda_0 \neq 1)$.

Contrasting to Huang's technique depending on Ridge Regression to append the threshold to each component of matrix $H^T H(\hat{\beta} = (H^T H + 1/C)^{-1} H^T T)$, the projected R-ELM technique just set a proper threshold to some singular element to matrix D subsequent to LDL^T decomposition to HTH. In contradiction, the enhanced technique appears to be better realistic.

Asymmetric matrix is considered for further enhancement instead of symmetric matrix and this technique is proposed for classification of pedestrians termed as Asymmetric Least Squares RELM (ARELM). It has advantages over traditional ELM and at the same time it retains the speed of learning of ELM technique. The proposed technique can handle wide variety of data rather taking into the account of the expectile distance of the boundary than taking into account of the least value.

5 Experimental Results

Caltech Dataset is taken for carrying out the analysis by using the suggested classification technique. Functional analysis is performed under the six conditions using the Caltech Pedestrian Dataset. The six different conditions listed are:

(i) Overall, (ii) Near Scale, (iii) Medium Scale, (iv) No Occlusion, (v) Partial Occlusion and (vi) Reasonable Occlusion. Rate of Miss versus False Positives were evidently visualized from Figs. 3, 4, 5, 6, 7 and 8 and log average rate of miss will be taken into account as usual benchmark value.

Overall

From Fig. 4 it is observed that the evaluation result to detect person on road using Frame Level Features with the proposed ARELM outperforms the other classification strategies and accomplished the log average miss rated 78% which is far better than penultimate Multi-Filter motion which is 81% for entire annotated pedestrian dataset.

Scale

Scale condition has taken based on the size of the pedestrian, calculated using height in pixel. In this, experimental results are in accordance with the nearer and medium scale under the circumstance of unblocked individuals on the road. Here medium-scale pedestrian has height range from 30 to 80 pixels and far scale pedestrian has pixels less than or equal to 20. For the Near Scale Frame Level Features with ARELM shows the superior functioning ability with log average miss rate of 51% by comparison with other detectors that accomplish the miss rate of 60–80% which are presented in Fig. 5. With Medium Scale, functional accuracy using Frame Level features with ARELM Classification strategy falls to the log average miss rate of 55%. On other hand it outperforms the other indicator employed with the miss rate of 60–90% that might be visibly observed in Fig. 6.

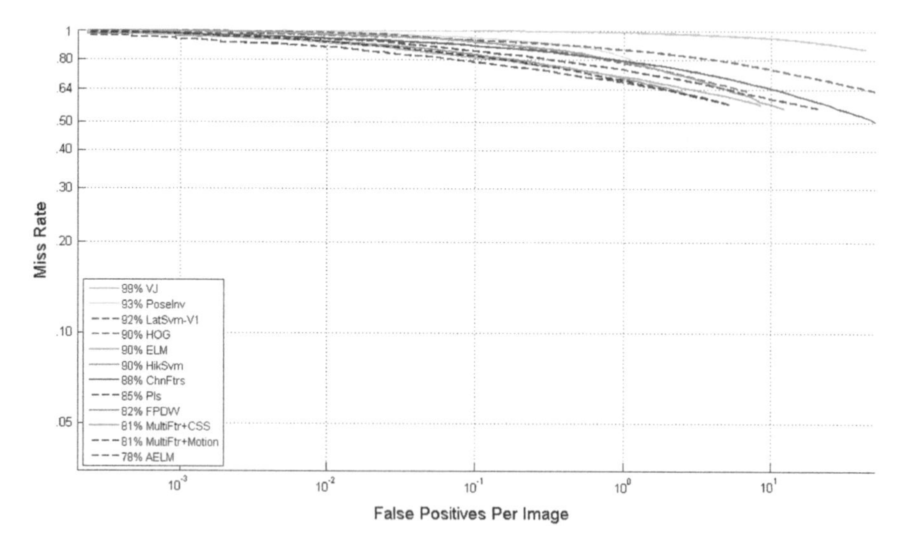

Fig. 4 Evaluation result of entire annotated pedestrian

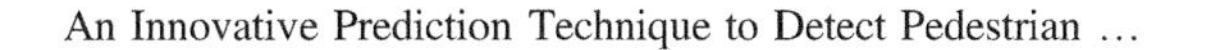

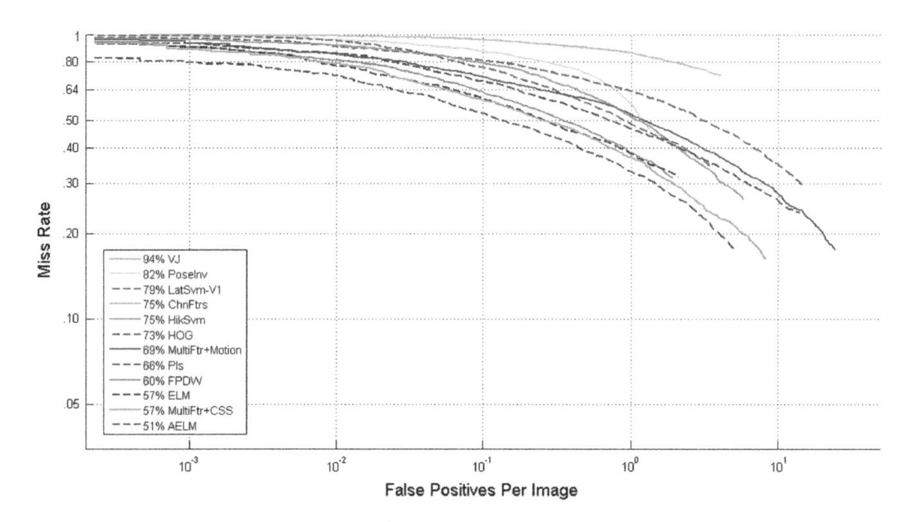

Fig. 5 Near scale pedestrian

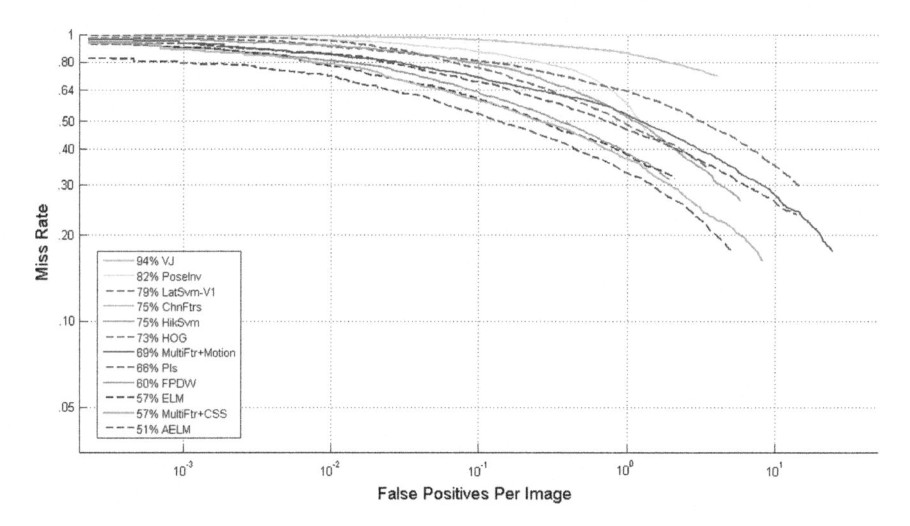

Fig. 6 Medium scale pedestrian

Occlusions

Classification of pedestrians done based on the unblocked, partial blocked (1–35% occluded) and profoundly blocked (35–80% occluded) conditions. Accuracy in the rate of detection drips considerably though with fractional blockage, resulting in the log-average miss rate of 51% for ARELM is noticed in Fig. 7. In the case of no occlusion and in the circumstance of partial occlusion accomplishment of log average miss rate of 77% using proposed technique is observed with respect to Fig. 8.

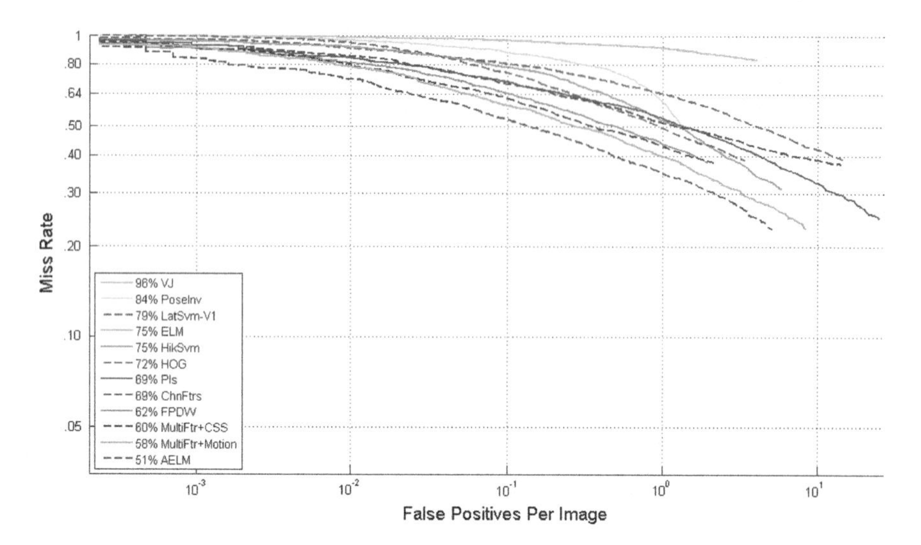

Fig. 7 Pedestrian without occlusion

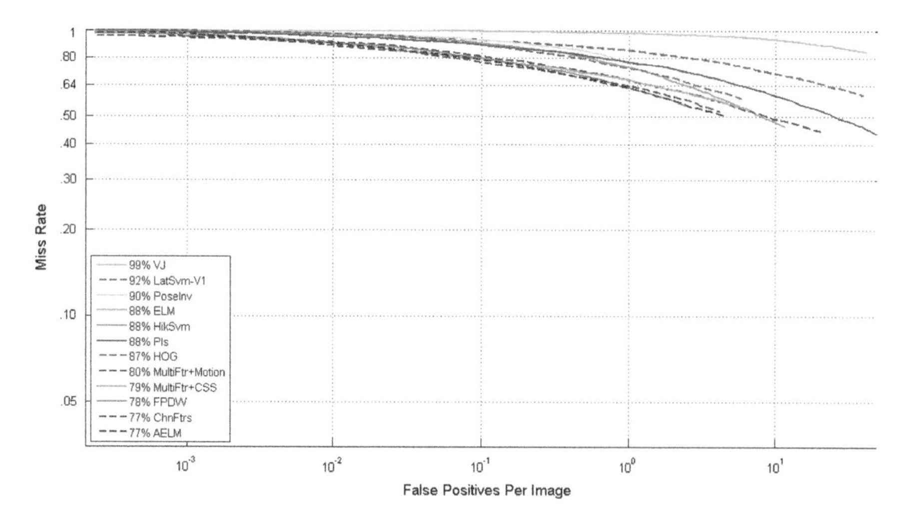

Fig. 8 Pedestrian in partial occlusion

Reasonable

Falling in the accuracy of recognition inspired the enquiry in performing the researches with the recognition of pedestrians with the minimum of 60 Pixels tall with the circumstance of no blockage or partial blockage. From Fig. 9 it could be undoubtedly observed that ARELM with Frame Level Feature delivers the upgraded functionality with better accuracy containing the log average miss rate of 87% by performing the analysis with the other detectors.

Runtime Analysis

With numerous areas such as pedestrian recognition, together with computerized protection, supervision, robotics, and person machine communications, time taken to recognize will be considered as the significant factor. Even though during the entire analysis, the research work concentrated on correctness; and also, it is deliberated that with the help of the accuracy and rapidity. Moreover, because in real time circumstance it is considered that the duration of detection will also significantly contribute towards the safety of the pedestrian.

From Fig. 10 it is observed that the detecting time for pedestrian using the suggested ARELM classification technique is extremely fast when compared with the other classification techniques. It is noticed that with ARELM the detection rate

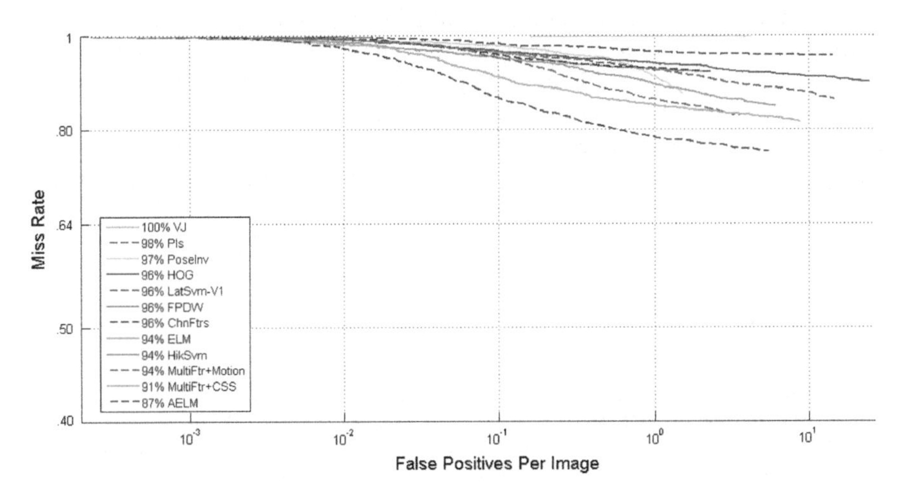

Fig. 9 Under the condition of reasonable occlusion

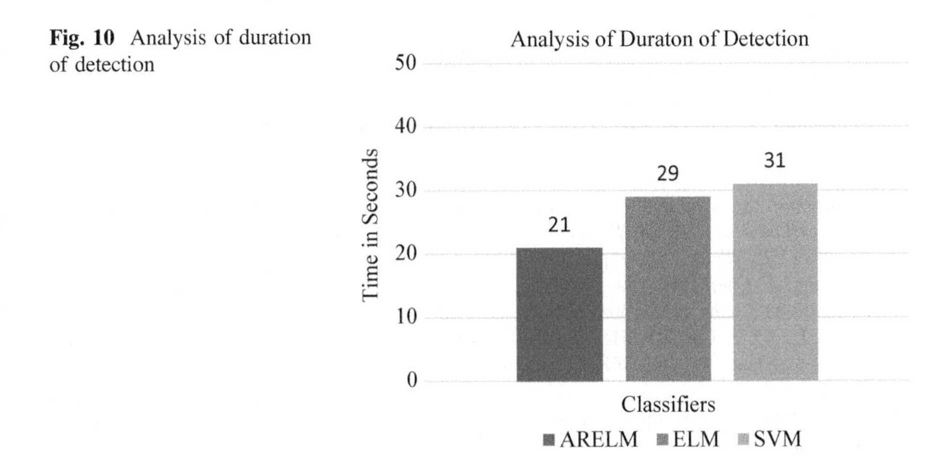

Fig. 10 Analysis of duration of detection

is faster because of the inherent nature of Extreme Learning Machine that contains the fastest learning speed. But here Extreme Learning Machine is improved in the suggested technique using Asymmetric Least Squares R-Extreme Learning Machine. Therefore, apart from improving the accuracy of detection rate by avoiding the false positives, it improves the speed of detection too.

6 Conclusion

Since Pedestrian recognition is the prominent issue in the automatic visualization system, Machine learning proves to be a competent strategy for the purpose of recognizing the substances. This paper describes a probabilistic framework for pedestrian detection with occlusion handling by increasing the accuracy and also accelerating the duration of pedestrian recognition. An improved technique of ARELM strategy is utilized in predicting the presence of person on road with the improved accuracy by reducing the false detection rate. Experimental results demonstrate that proposed ARELM has the highest accuracy and least quantity of false positives with high detection speed when compared to other classifiers used in the detection of pedestrians.

References

1. W. Kim, S. Lee, Pedestrian Detection Using Structured SVM (2013)
2. M. Higashikubo, Y. Ogiuchi, Y. Ono, T. Kurita, K. Nishida, H. Inayoshi, R. Arata, Measurement of vehicles and motorcycles with detection by support vector machines and pair-pixel feature tracking, in *Symposium on Sensing via Image Information (SSII)*, CD-R, IS4-02 (2010)
3. F. Ren, J. Huang, M. Terauchi, R. Jiang, R. Klette, Lane detection on the iPhone, in *International Conference on Arts and Technology* (Springer, Berlin, Heidelberg, 2009), pp. 198–205
4. M. Takagi, H. Fujiyoshi, Road sign recognition using SIFT feature, in *Symposium on Sensing via Image Information (SSII)*, CD-R, LD2-06 (2007)
5. K. Matsushima, Z. Hu, K. Uchimura, Pedestrian recognition using stereo sensor, in *Information Processing Society of Japan (IPSJ), SIG Technical Reports ITS*, pp. 49–54 (2006)
6. Y. Nakashima, J.K. Tan, H. Kim, S. Ishikawa, A pedestrian detection method using the extension of the HOG feature, in *2014 Joint 7th International Conference on Soft Computing and Intelligent Systems (SCIS) and 15th International Symposium on Advanced Intelligent Systems (ISIS)* (IEEE, 2014), pp. 1198–1202
7. S. Wang, J. Cheng, H. Liu, M. Tang, Pcn: part and context information for pedestrian detection with cnns. arXiv preprint arXiv:1804.04483 (2018)
8. Y. Xu, D. Xu, S. Lin, T.X. Han, X. Cao, X. Li, Detection of sudden pedestrian crossings for driving assistance systems. IEEE Trans. Syst. Man Cybern. Part B (Cybern.) **42**(3), 729–739 (2012)
9. P.J. Phillips, Human identification technical challenges, in *Proceedings. International Conference on Image Processing*, vol. 1 (IEEE, 2002), p. I

10. O.L. Junior, D. Delgado, V. Gonçalves, U. Nunes, Trainable classifier-fusion schemes: an application to pedestrian detection, in *12th International IEEE Conference on Intelligent Transportation Systems* (IEEE, 2009), pp. 1–6
11. N. AlDahoul, M. Sabri, A. Qalid, A.M. Mansoor, Real-time human detection for aerial captured video sequences via deep models. Comput. Intell. Neurosci. (2018)
12. N. Dalal, B. Triggs, Histograms of oriented gradients for human detection, in *International Conference on Computer Vision & Pattern Recognition (CVPR'05)*, vol. 1 (IEEE Computer Society, 2005), pp. 886–893
13. P. Viola, M.J. Jones, D. Snow, Detecting pedestrians using patterns of motion and appearance. Int. J. Comput. Vision **63**(2), 153–161 (2005)
14. A. Vedaldi, V. Gulshan, M. Varma, A. Zisserman, Multiple kernels for object detection, in *IEEE 12th International Conference on Computer Vision* (IEEE, 2009), pp. 606–613
15. O. Tuzel, F. Porikli, P. Meer, Pedestrian detection via classification on riemannian manifolds. IEEE Trans. Pattern Anal. Mach. Intell. **30**(10), 1713–1727 (2008)
16. D. Hoiem, A.A. Efros, M. Hebert, Putting objects in perspective. Int. J. Comput. Vision **80**(1), 3–15 (2008)
17. P. Sermanet, K. Kavukcuoglu, S. Chintala, Y. LeCun, Pedestrian detection with unsupervised multi-stage feature learning, in *Proceedings of the IEEE Conference on Computer Vision and Pattern Recognition*, pp. 3626–3633 (2013)
18. W. Ouyang, X. Wang, Joint deep learning for pedestrian detection, in *Proceedings of the IEEE International Conference on Computer Vision*, pp. 2056–2063 (2013)
19. J.J. Lim, C.L. Zitnick, P. Dollár, Sketch tokens: a learned mid-level representation for contour and object detection, in *Proceedings of the IEEE Conference on Computer Vision and Pattern Recognition*, pp. 3158–3165 (2013)
20. L. Wang, B. Zhang, Boosting-like deep learning for pedestrian detection. arXiv preprint arXiv:1505.06800 (2015)
21. H. Li, Z. Lin, X. Shen, J. Brandt, G. Hua, A convolutional neural network cascade for face detection, in *Proceedings of the IEEE Conference on Computer Vision and Pattern Recognition*, pp. 5325–5334 (2015)
22. J. Cao, Y. Pang, X. Li, Learning multilayer channel features for pedestrian detection. IEEE Trans. Image Process. **26**(7), 3210–3220 (2017)
23. H. Ameur, A. Helali, M. Nasri, H. Maaref, A. Youssef, Improved feature extraction method based on histogram of oriented gradients for pedestrian detection, in *2014 Global Summit on Computer & Information Technology (GSCIT)* (IEEE, 2014), pp. 1–5
24. H. Li, Z. Wu, J. Zhang, Pedestrian detection based on deep learning model, in *2016 9th International Congress on Image and Signal Processing, BioMedical Engineering and Informatics (CISP-BMEI)* (IEEE, 2016), pp. 796–800
25. M. He, H. Luo, Z. Chang, B. Hui, Pedestrian detection with semantic regions of interest. Sensors **17**(11), 2699 (2017)
26. M. Bilal, M.S. Hanif, High performance real-time pedestrian detection using light weight features and fast cascaded kernel SVM classification. J. Signal Process. Syst. **91**(2), 117–129 (2019)
27. M.T.T. Nguyen, V.D. Nguyen, J.W. Jeon, Real-time pedestrian detection using a support vector machine and stixel information, in *17th International Conference on Control, Automation and Systems (ICCAS)* (IEEE, 2017), pp. 1350–1355
28. H.S.G. Supreeth, C.M. Patil, An adaptive SVM technique for object tracking. Int. J. Pure Appl. Math. **118**(7), 131–135 (2018)
29. W. Zheng, S. Cao, X. Jin, S. Mo, H. Gao, Y. Qu, W. Jiang, Deep forest with local experts based on elm for pedestrian detection, in *Pacific Rim Conference on Multimedia* (Springer, Cham, 2018), pp. 803–814
30. P. Navarro, C. Fernandez, R. Borraz, D. Alonso, A machine learning approach to pedestrian detection for autonomous vehicles using high-definition 3D range data. Sensors **17**(1), 18 (2017)

31. W. Ouyang, H. Zhou, H. Li, Q. Li, J. Yan, X. Wang, Jointly learning deep features, deformable parts, occlusion and classification for pedestrian detection. IEEE Trans. Pattern Anal. Mach. Intell. **40**(8), 1874–1887 (2018)
32. M. Errami, M. Rziza, Improving pedestrian detection using support vector regression, in *13th International Conference on Computer Graphics, Imaging and Visualization (CGiV)* (IEEE, 2016), pp. 156–160
33. B. Wang, S. Wang, X. Liu, J. Yang, Effective object tracking using extreme learning machine with smoothness and preference regularization. Electron. Lett. **51**(23), 1867–1869 (2015)
34. Y. Yu, L. Xie, Z. Huang, An object tracking method using extreme learning machine with online learning, in *2016 IEEE Symposium Series on Computational Intelligence (SSCI)* (IEEE, 2016), pp. 1–7
35. R.A. Kharjul, V.K. Tungar, Y.P. Kulkarni, S.K. Upadhyay, R. Shirsath, Real-time pedestrian detection using SVM and AdaBoost, in *2015 International Conference on Energy Systems and Applications* (IEEE, 2015), pp. 740–743
36. C. Yang, H. Liu, S. Liao, S. Wang, Pedestrian detection in thermal infrared image using extreme learning machine, in *Proceedings of ELM-2014*, vol. 2 (Springer, Cham, 2015), pp. 31–40
37. A. Sumi, T. Santha, Motion deblurring for pedestrian crossing detection in advanced driver assistance system, in *2017 IEEE International Conference on Computational Intelligence and Computing Research (ICCIC)*, pp. 1–4 (2017)
38. M. Kazubek, Wavelet domain image denoising by thresholding and Wiener filtering. IEEE Signal Process. Lett. **10**(11), 324–326 (2003)
39. S. Chen, X. Wang, Y. Tang, X. Chen, Z. Wu, Y.G. Jiang, Aggregating frame-level features for large-scale video classification. arXiv preprint arXiv:1707.00803 (2017)
40. H.G. Zhang, S. Zhang, Y.X. Yin, A novel improved ELM algorithm for a real industrial application. Math. Probl. Eng. (2014)

Artificial Intelligence Applications in Tracking Health Behaviors During Disease Epidemics

Kurubaran Ganasegeran and Surajudeen Abiola Abdulrahman

Abstract The threat of emerging and re-emerging infectious diseases to global population health remains significantly enormous, and the pandemic preparedness capabilities necessary to confront such threats must be of greater potency. Artificial Intelligence (AI) offers new hope in not only effectively pre-empting, preventing and combating the threats of infectious disease epidemics, but also facilitating the understanding of health-seeking behaviors and public emotions during epidemics. From a systems-thinking perspective, and in today's world of seamless boundaries and global interconnectivity, AI offers enormous potential for public health practitioners and policy makers to revolutionize healthcare and population health through focussed, context-specific interventions that promote cost-savings on therapeutic care, expand access to health information and services, and enhance individual responsibility for their health and well-being. This chapter systematically appraises the dawn of AI technology towards empowering population health to combat the rise of infectious disease epidemics.

Keywords Artificial intelligence · Health behaviors · Epidemics · Infectious disease · Global health

K. Ganasegeran (✉)
Clinical Research Center, Seberang Jaya Hospital, Ministry of Health Malaysia, Penang, Malaysia
e-mail: medkuru@yahoo.com

S. A. Abdulrahman
Emergency Medicine Department, James Paget University Hospital, Great Yarmouth, Norfolk, UK
e-mail: abdulsuraj@gmail.com

D. J. Hemanth (ed.), *Human Behaviour Analysis Using Intelligent Systems*, Learning and Analytics in Intelligent Systems 6,
https://doi.org/10.1007/978-3-030-35139-7_7

1 Introduction

Infectious diseases disrespect national and international borders. They pose substantial threats and serious repercussions to global public health security. While the Asia-Pacific region was generally regarded as the main epicenter of emerging infectious diseases, with outbreaks of Avian Flu, Asian Flu and Severe Acute Respiratory Syndrome (SARS) [1], the recent and unexpected emergence of Zika pandemic spurred global concerns about pandemic preparedness capabilities particularly as it relates to training and deployment of healthcare workforce at a massive level, worldwide. Despite coordinated global efforts, containing the "red alert" pandemic of Zika remained a challenge, as both healthcare workers and public health advocates were uncertain about such disastrous contagion causing serious complications including congenital microcephaly in newborns and neurological deficits in adults [2, 3].

Control measures were obtunded as public health advocates were initially speculative about the potential transmission route of Zika, while clinicians in hospitals were irresolute, instituting multiple levels of care and management to tackle the complications of Zika. This debacle gave rise to an urgent need to debate the circumstances under which the Zika epidemic has challenged human intelligence behavior and capacity to battle the threat effectively and efficiently.

As population explosions and uncontrolled human mobility across nations catalyzes rapid disease propagation, our next question is, what else above human intelligence could help resolve such unprecedented epidemic crisis? Scientists believe that the time has come to institute analytic technologies—such as Artificial Intelligence (AI)—in healthcare to help prevent and resolve such large disease epidemics [4, 5]. Adaptive AI applications could mould human behavior to practice preventive behaviors and disease control strategies [6], thereby improving global health. This chapter will systematically discuss the dawn of AI technology in healthcare that could potentially empower the human population to tackle unprecedented infectious disease epidemics.

2 Artificial Intelligence—The Evolution Begins!

The human population has witnessed four major revolutions till date (Fig. 1); the foremost being the first industrial revolution that introduced steam engine to the world [7]. This was followed by the second industrial revolution that introduced electrical-energy based productions. The first information revolution was conceptualized during the third industrial revolution in the late 20th century. It was during this time that computers and internet-based knowledge began and has since then shaped human interactions. In early 21st century, the fourth industrial revolution accelerated the second information revolution. The entire phase of human daily functions transformed with the debut of AI, bringing together massive information

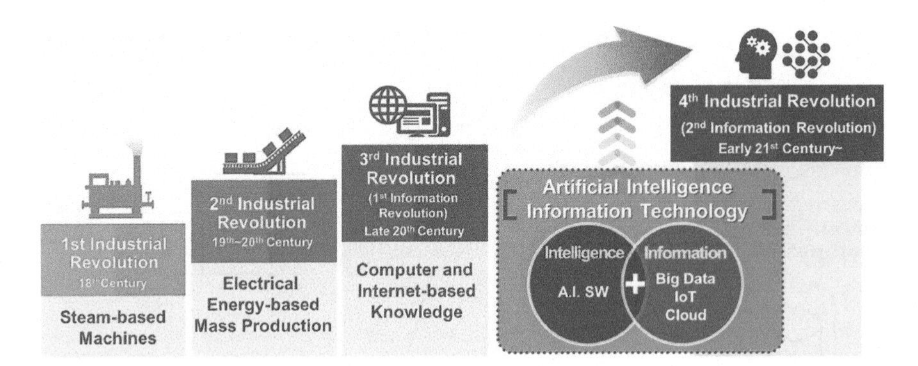

Fig. 1 The four industrial revolutions within the human interface. *Source* http://cradall.org/content/being-human-era-fourth-industrial-revolution-and-ai-peter-kearns with original source from worldbank.org. Figure available in public domain with licensee agreement of "*fair use*," allowing free usage for academic and research purposes

flow from different specialties. These culminated in the rise of Big Data with systems integration across the Internet of Things (IOT) and Cloud Computing Systems.

Current revolutionary era is based on extreme automation for global connectivity, in which AI would definitely play an imperative role as a resource to utilize. At the peak of emergent multi-function contexts of AI and the rise of Big Data Analytics, the United Nations (UN) in 2017 unified global experts to galvanize a dynamic consensus on the adoption and expansion of AI use in delivering good public care services [8]. Succinctly, various stakeholders were assembled together in another UN meeting to assess the role of AI towards achieving Sustainable Developmental Goals (SDGs) [9]. From the healthcare perspective, massive data have been obtained from public health surveillance efforts with the advancement of AI. One major public health field that gained momentum to develop various AI applications for disease prevention was the infectious disease domain [5]. The human population is currently able to access potentially useful massive data sources of infectious disease spread through sentinel reporting systems, national surveillance systems (usually operated by national or regional disease centers such as the Center for Disease Control (CDC)), genome databases, internet search queries (also called infodemiology and infoveillance studies) [10–12], Twitter data analysis [13, 14], outbreak investigation reports, transportation dynamics [15], vaccine reports [16] and human dynamics information [17].

With the influx of massive data volume, effective data integration, management and knowledge extraction systems are required [5]. Epidemic modeling and disease-spread simulations form new horizons to understand the effects of citizen behaviors or government health policy measures [1]. A simple integrated effect of disease knowledge discovery is exhibited in Fig. 2.

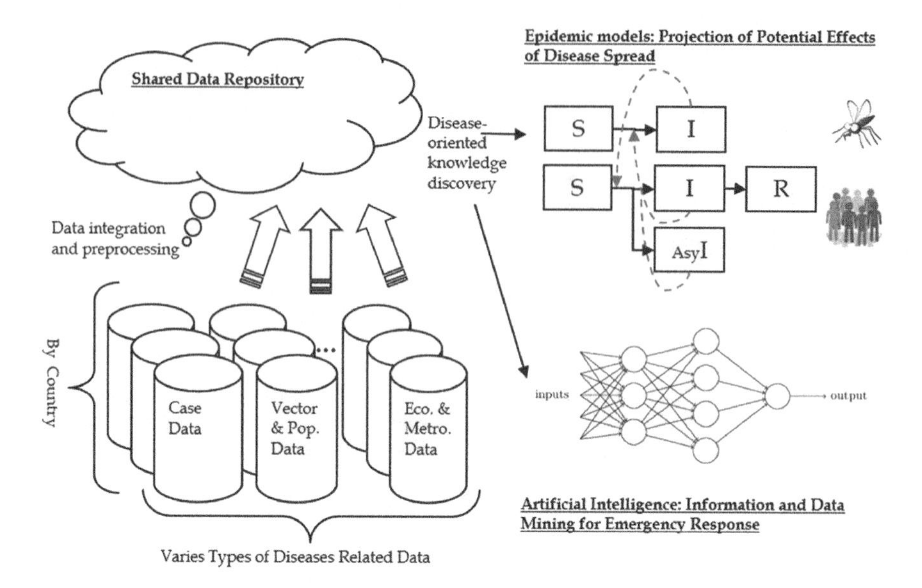

Fig. 2 Integrated conceptual model for infectious diseases using AI. *Source* Wong et al. [5]

3 Conceptualizing Artificial Intelligence in Disease Epidemics

As humans, we are able to perform simple essential tasks such as object detection, visual interpretation and speech recognition.

Our interpretation is instantaneous when we look at an object or image, or when we hear voices or noises surrounding us. Our next question is—could AI perform these essential intellectual tasks as well? The answer is absolutely yes, but in a different mode of function. While human interpretation is solely dependent on cognitive functions, AI requires mathematical algorithms to automate machines for execution of such functions [18]. Machines here refer to programmable computers! An example is to visualize the cause of an outbreak; Dengue, Chikungunya or Zika, of which these diseases are commonly caused by the vector mosquito. In massive epidemics, elimination of the vector is important, and human cognitive functions can never detect all mosquitos in an outbreak investigation area! However, this can be easily detected through deployment of AI in areas which have loads of mosquito vectors to facilitate control measures. Figure 3 exhibits how human and AI technology interpret the vector differently.

While human interpretation is instantaneous, AI evaluates the same image as humans do, but translated into codes [18], facilitating massive detection. While AI aims to mimic human cognitive functions, it lacks intuitive behaviors. Scientists postulate that such synthetic intelligence which could be on par with human intelligence can be called "*computational intelligence*." However, the primary goal

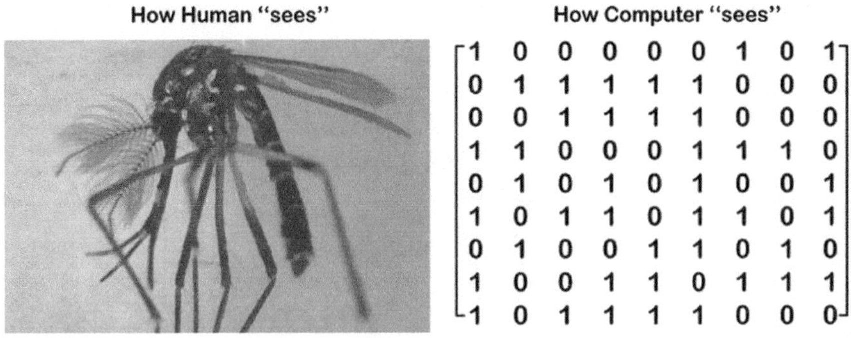

How Human "sees"

How Computer "sees"

$$\begin{bmatrix} 1 & 0 & 0 & 0 & 0 & 0 & 1 & 0 & 1 \\ 0 & 1 & 1 & 1 & 1 & 1 & 0 & 0 & 0 \\ 0 & 0 & 1 & 1 & 1 & 1 & 0 & 0 & 0 \\ 1 & 1 & 0 & 0 & 0 & 1 & 1 & 1 & 0 \\ 0 & 1 & 0 & 1 & 0 & 1 & 0 & 0 & 1 \\ 1 & 0 & 1 & 1 & 0 & 1 & 1 & 0 & 1 \\ 0 & 1 & 0 & 0 & 1 & 1 & 0 & 1 & 0 \\ 1 & 0 & 0 & 1 & 1 & 0 & 1 & 1 & 1 \\ 1 & 0 & 1 & 1 & 1 & 1 & 0 & 0 & 0 \end{bmatrix}$$

Fig. 3 Interpreting the vector from the human and AI perspective. *Source* da Silva Motta et al. [18]

of AI was to create a system programming that is capable to think and act rationally like humans, although such machines may lack intuitive or emotional capabilities. As such, AI has been appropriately defined in simple and straightforward terms, as "*a branch of computer science that deals with simulation of intelligent behaviors as humans using computers* [19]".

4 Types of AI

In principle, there are three types of AI. If a machine is able to think as humans do and perform a task similar to human intellectual capabilities, then that machine functionality is referred to as artificial general intelligence [20]. If a machine performs a single task extremely well, this is known as artificial narrow intelligence [4]. If the same machine out-smart the best humans in all fields from scientific creativity to general wisdom or social skills, this is referred to as artificial super intelligence [4]. At present, virtually all contemporary AI application systems utilize artificial narrow intelligence.

4.1 Main Subsets of AI

There are numerous concepts to function underlying AI applications in healthcare. Based on the required functions, these concepts are clumped together to automate a single application—such as tracking infectious disease health seeking behavior. The following sub-sections summarize key concepts of different AI subsets adopted in emerging literature of infectious diseases.

Machine Learning (ML). ML is a subfield of AI that implies learning from previous experiences (Fig. 4). The system finds solution to a problem by extracting

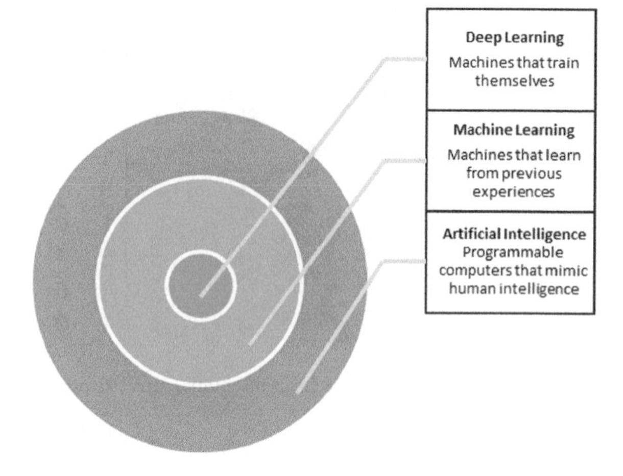

Fig. 4 Diagram Venn illustrating artificial intelligence and main subsets

previous relevant data, learn from this data and predicts new outcomes [20]. ML applications are sub-divided into three categories:

I. *Supervised learning:* uses patterns of identified data (e.g. training data)
II. *Unsupervised learning:* finds and learns from patterns of data (e.g. data-mining that involves identification of patterns in large datasets)
III. *Reinforcement learning:* an extension of supervised learning that "rewards" and "punishes" when an application interacts with the environment.

Table 1 illustrates some common examples of supervised and unsupervised ML methods that are currently adopted and utilized to track health seeking behaviors during infectious disease epidemics [5].

Deep Learning (DL). DL is a specific subset of ML that uses neural networks (Fig. 4). In short, it is basically a synthetic replica of the human brain structure and functionality [4]. DL can execute multiple functions like image recognition and natural language processing (NLP). The system is capable of handling large datasets of information flow.

Image Recognition. AI has the capability to process large amount of data about characteristics of a particular phenomenon in the form of images or signals [20]. Motion images and sounds are examples of signals that could be analyzed using artificial neural networks (ANNs) [20]. Recently, researchers from the USA proposed a system that could rapidly identify potential arbovirus outbreaks (mosquito, ticks or other arthropod borne viruses) [21]. The system identifies images of mosquito larvae captured and delivered by a group of citizen scientists. Not only did the developed prototype facilitate collection of images, it also facilitated training of image classifiers for the recognition of a particular specimen. This sets a base for execution of expert validation process and data analytics. It was found that recognition of specimen in images provided by citizen scientists was useful to generate visualizations of susceptible geographical regions of arboviruses threat (Fig. 5). The system was capable of identifying mosquito larvae with great accuracy.

Table 1 Common ML methods applicable to track health-behaviors during infectious disease epidemics

Types	Examples of methods used	Functions
Supervised method	Support Vector Machine (SVM), Decision Tree, Random Forest, Naive Bayes (NB), Artificial Neural Network (ANN), Bootstrap Aggregating, AdaBoost	The primary function of these methods is to create prediction outcomes to warn authorities and public in advance on potential infectious disease epidemics, thus suggesting immediate prevention and control strategies
Unsupervised method	Principal Component Analysis (PCA)	This method aims to reduce data dimensions to enable researchers easily uncover key factors causing infectious disease dynamics
	K-Means	This method clusters patients for abnormality detections
	Latent Dirichlet Allocation (LDA)	This method enables data extraction from medical contextual records
	Deep Learning Architectures	This method facilitates prediction and classification, social network filtering and applications of bioinformatics in infectious disease analytics
	Structural Equation Model Trees	This method allows estimation of complex cause-effect relationship models with latent variables

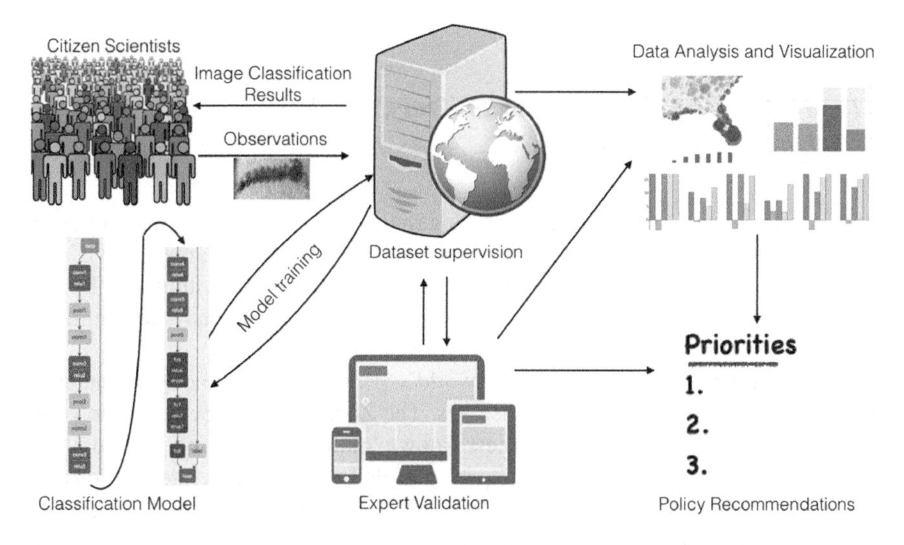

Fig. 5 Proposed system for rapid identification of arbovirus outbreaks. *Source* Munoz et al. [21]

The rapid identification of potential outbreak to a susceptible community could alert preventive behaviors and policy drafting in the quest to control potential epidemics.

Natural Language Processing (NLP). NLP bridges the gap between languages that humans and machines use to operate. Algorithms are built to allow machines to identify keywords and phrases in an unstructured written text. AI applications then interpret the meaning of these texts for actionable knowledge [20].

Expert Systems (ES). ES incorporates expert-level competence to resolve a particular problem [20]. The system is constituted of two main components, namely knowledge base and a reasoning engine. It solves complex problems through reasoning a set of incomplete or uncertain information through a series of complex rules. In recent years, fuzzy logic, a set of mathematical principles for knowledge representation was crafted to accelerate the evolution of ES. Such strategy was utilized by a team of researchers from South Africa to improve predictions of cholera outbreaks [22].

5 Applications of AI During Disease Epidemics: Concept Examples

Public reaction and behavior towards disease outbreaks could be difficult to predict. With the rise of Big Data Analytics and a pool of AI applications in place, public health researchers were able to correlate population's behavior during an outbreak [5]. The following examples illustrate real life applications of AI during disease epidemics:

5.1 The Use of Twitter Data to Track Public Behavior During Outbreak Crisis

Twitter, a free social-networking micro-blogging service has enabled loads of users to send and read each other's "tweets (short, 140-character messages)." As important information and geo-political events are embedded within the Twitter stream, researchers now postulate that Twitter users' reactions may be useful for tracking and forecasting behavior during disease epidemics.

The Zika Pandemic. Most of the world's populations are living in endemic areas for common mosquito-borne diseases. The Zika pandemic between the years of 2015 and 2016 marked the largest known outbreak, reaching a "red-alert" warning of multiple complications requiring global public health interventions. In such exigencies, population health behaviors are important for potential control measures. Daughton and Paul postulated that internet data has been effective to track human health seeking behaviors during disease outbreaks [12]. They used

Twitter data between 2015 and 2016 respectively to identify and describe self-disclosures of an individual's behavior change during disease spread. They combined keyword filtering and ML classifications to identify first-person reactions to Zika. A total of 29,386 English tweets were analyzed. Keywords include "travel," "travelling intentions," and "cancellations." Individual demographic characteristics, users' networking and linguistic patterns were compared with controls. The study found variations between individual characteristics, users' social network attributes and language styles in 1567 Twitter users. These users changed or considered to change their travel behaviors in response to Zika. Significant differences were observed between geographic areas in the USA, with higher discussion among women than men and some differences in levels of exposure to Zika-related information. This finding concludes that applying AI concepts could contribute to better understanding on how public perceives and reacts towards the risks of infectious disease outbreak.

The Influenza A H1N1 Pandemic. Signorini and colleagues in 2011 analyzed Twitter embedded data for tracking rapid evolvement of public concerns with respect to H1N1 or swine flu, while concurrently measuring actual disease activity [13]. The researchers explored public concerns by collecting tweets using pre-specified search terms related to H1N1 activity with additional keywords related to disease transmission, disease countermeasures and food consumption within the United States. They utilized influenza-like illness surveillance data and predicted an estimation model using supervised learning method in machine learning. The results showed that Twitter was useful to measure public interest or concern about health-related events associated with H1N1. These include an observed periodical spikes related to user Twitter activity that were linked to preventive measures (hand-hygiene practices and usage of masks), travel and food consumption behaviors, drug related tweets about specific anti-viral and vaccine uptake. They concluded that Twitter accurately estimated influenza outbreak through AI applications [13].

5.2 Infodemiology and Infoveillance Studies to Track Health-Behaviors

The integration of internet data into public health informatics has been regarded as a powerful tool to explore real-time human health-seeking behaviors during disease epidemics. One such popular tool widely utilized is Google Trends, an open tool that provides traffic information regarding trends, patterns and variations of online interests using user-specified keywords and topics over time [11]. Such adaptations formed two conceptualizations: the first was "*Infodemiology*," defined as "*the science of distribution and determinants of information in an electronic medium, specifically the Internet, or in a population, with the ultimate aim to inform public health and public policy* [23];" the second was "*Infoveillance*," defined as "*the longitudinal tracking of infodemiology metrics for surveillance and trend analysis* [24]."

Examining Health-Behavior Patterns during Dengue Outbreaks. Dengue is highly endemic across the South-East Asian countries. Recently, a group of researchers from the Philippines conducted an infodemiology and infoveillance study by using spatio-temporal concepts to explore relationships of weekly Google Dengue Trends (GDT) data from the internet and dengue incidence data from Manila city between 2009 and 2014 [25]. They subsequently examined health-seeking behaviors using dengue-related search queries from the population. Their findings suggested that weekly temporal GDT patterns were nearly similar to weekly dengue incidence reports. Themes retrieved from dengue-related search queries include: "dengue," "symptoms and signs of dengue," "treatment and prevention of dengue," "mosquito," and "other diseases." Most search queries were directed towards manifestations of dengue. The researchers concluded that GDT is a useful component to complement conventional disease surveillance methods. This concept could assists towards identifying dengue hotspots to facilitate appropriate and timely public health decisions and preventive strategies [25].

Health-Seeking Behavior of Ebola Outbreak. An unprecedented Ebola contagion that plagued most West African countries in 2014 marked the rise of global public health interest in pandemic preparedness interventions. Millions of Ebola-related internet hit searches were retrieved. With such high fluxes of health-seeking behavior using computers, a group of Italian researchers' evaluated Google Trends search queries for terms related to "Ebola" outbreak at the global level and across countries where primary cases of Ebola were reported [26]. The researchers subsequently explored correlations between overall and weekly web hit searches of terms in relation to the total number and weekly new cases of Ebola incidence. The highest search volumes that generated Ebola related queries were captured across the West African countries, mainly affected by the Ebola epidemic. Web searches were concentrated across state capitals. However, in Western countries, the distribution of web searches remained fixed across national territories. Correlations between the total number of new weekly cases of Ebola and the weekly Google Trends index varied from weak to moderate among the African countries afflicted by Ebola. Correlations between the total number of Ebola cases registered in all countries and the Google Trends index was relatively high. The researchers concluded that Google Trends data strongly correlated with global epidemiological data. Global agencies could utilize such information to correctly identify outbreaks, and craft appropriate actionable interventions for disease prevention urgently [26].

Public Reactions toward Chikungunya Outbreaks. The 2017 Italian outbreak of Chikungunya posed substantial public health concerns, catalyzing public interests in terms of internet searches and social media interactions. A group of researchers were determined to investigate Chikungunya-related digital health-seeking behaviors, and subsequently explored probable associations between epidemiological data and internet traffic sources [27]. Public reactions from Italy toward Chikungunya outbreaks were mined from Google Trends, Google News, Twitter traffics, Wikipedia visits and edits, and PubMed articles to yield a structural equation model. The relationships between overall Chikungunya cases, as well as autochthonous cases and tweet productions were mediated by Chikungunya-related web searches. But in the allochthonous case model, tweet productions were not significantly mediated by epidemiological figures, instead, web searches posed significant mediating tweets. Inconsistent associations were detected in mediation models involving Wikipedia usage. The effects between news consumption and tweets production were suppressed in this regard. Subsequently, inconsistent mediation effects were found between Wikipedia usage and tweets production, with web searches as a mediator. After adjustment of Internet Penetration Index, similar findings were retrieved with the adjusted model showing relationship between Google News and Twitter to be partially mediated by Wikipedia usage. The link between Wikipedia usage and PubMed/MEDLINE was fully mediated by Google News, and differed from the unadjusted model. The researchers found significant public reactions to the Chikungunya outbreak. They concluded that health authorities could be made aware immediately of such phenomenon with the aid of new technologies for collecting public concerns, disseminating awareness and avoiding misleading information [27].

5.3 Computer Based Expert Systems

Expert systems are built upon the basis to act as a diagnostic tool to accelerate detection of infectious disease epidemics, determining the intensity or concentration of vector-agents within the triads of infectious disease dynamics.

The Malaria Control Strategy Using Expert Systems. Malaria constitutes a "red-alert" health threat to the African communities. A group of researchers from Nigeria built an expert system for malaria environmental diagnosis with the aim of providing a decisional support tool for researchers and health policy-makers [28]. As prevailing malaria control measures were deemed insufficient, this group of researchers developed a prototype that constituted components of "knowledge," "applications," "system database," "user graphics interface," and "user components." The user component utilized Java, while the application component used Java Expert System Shell (JESS) and the Java IDE of Netbeans. The database component used SQL Server. The system was able to act as a diagnostic tool to determine the intensity of malarial parasites in designated geographical areas across Africa. The proposed prototype proved useful and cost-effective in curbing malaria spread [28].

6 Challenges of AI Applications During Disease Epidemics

Whereas AI is gaining increasing popularity and acceptance as a quick fix to the myriad of challenges faced with pandemic preparedness using traditional population-based approaches, it is not without its own limitations. Even in resource-rich settings, there are challenges associated with building and updating the knowledge base of expert systems [29], providing high-quality datasets upon which machine learning algorithms can be premised, and ethical issues associated with data ownership and management [20]. Additionally, resource-limited settings are further plagued with constraints of poorly organized and integrated health systems, poor IT and communication infrastructure, and socio-economic and cultural contexts [30, 31] that significantly impact successful implementation of AI systems. Beyond these, the dynamics of human behavior and other environmental covariates (such as mass/social media, public emotions, public policy etc.) may not only influence the accuracy of epidemic disease modeling frameworks but also impact health seeking behavior during epidemics [32]. More than ever before, public health experts, IT developers and other stakeholders must work together to address concerns related to scalability of AI for healthcare, data integration and interoperability, security, privacy and ethics of aggregated digital data. Finally, the transparency of predictive AI algorithms have been called to question, particularly given their 'black box' nature which makes them prone to biases in settings of significant inequalities [33].

7 Opportunities and Future Direction of AI Applications During Disease Epidemics

Perhaps, it may be premature to describe AI as the future of healthcare given it is still in its infancy, however, it has become increasingly difficult to not acknowledge the substantial contributions of AI systems to the field of public health medicine. Notwithstanding current challenges with the widespread adoption of AI particularly in resource-limited settings, the use of AI in providing in-depth knowledge on individuals' health, predicting population health risks and improving pandemic preparedness capabilities is likely to increase substantially in the near future [34]. Further, the rapidly expanding mobile phone penetrance, developments in cloud computing, substantial investments in health informatics, electronic medical records (EMRs) and mobile health (mHealth) applications, even in resource-constrained settings, holds significant promise for increasing use and scalability of AI applications in improving public health outcomes [35]. Public health policy, practice and research will continue to benefit from the expanding framework of infodemiology and infoveillance in analyzing health information search, communication and publication behavior on the internet [23, 24]. Advances in cryptographic technologies—including block chain is likely to allay fears and concerns with security, privacy and confidentiality of public digital data/information [36].

8 Conclusion

There is no doubt that AI is and will continue to revolutionize healthcare and population health. From prevention and health promotion to diagnosis and treatment, AI is increasingly being deployed to improve clinical decision-making, enhance personalized care and public health outcomes. In particular, AI offers enormous potential for cost-savings on therapeutic care given its predictive accuracy of potential outbreaks and epidemics and ability to enhance positive health seeking behaviors (at individual and population levels) during epidemics predicated upon robust infodemiology and infoveillance frameworks supported by expert systems, machine learning algorithms and mobile applications. Amazing as the future of AI in healthcare seems, there are significant legal and ethical concerns that need to be addressed in order to pave way for robust implementation and scalability across a variety of socio-cultural, epidemiological, health system and political contexts.

Acknowledgements We thank the Ministry of Health Malaysia for the support to publish this chapter.

References

1. K.L. Tsui, Z.S.Y. Wong, D. Goldsman, M. Edesess, Tracking infectious disease spread for global pandemic containment. IEEE Intell. Syst. **28**(6), 60–64 (2013)
2. D. Baud, D.J. Gubler, B. Schaub, M.C. Lanteri, D. Musso, An update on Zika virus infection. Lancet **390**, 2099–2109 (2017)
3. I.R.F. da Silva, J.A. Frontera, A.M.B. de Filippis, O.J.M.D. Nascimento, RIO-GBS-ZIKV Research Group, Neurologic complications associated with the Zika virus in Brazilian adults. JAMA. Neurol. **74**(10), 1190–1198 (2017)
4. B. Mesko, G. Hetenyi, Z. Gyorffy, Will artificial intelligence solve the human crisis in healthcare? BMC Health. Serv. Res. **18**, 545 (2018)
5. Z.S.Y. Wong, J. Zhou, Q. Zhang, Artificial intelligence for infectious disease big data analytics. Infect. Dis. Health. **24**, 44–48 (2019)
6. S. Michie, J. Thomas, M. Johnston, P.M. Aonghusa, J. Shawe-Taylor, M.P. Kelly, L.A. Deleris, A.N. Finnerty, M.M. Marques, E. Norris, A. O'Mara-Eves, R. West, The human behavior-change project: harnessing the power of artificial intelligence and machine learning for evidence synthesis and interpretation. Implement. Sci. **12**, 121 (2017)
7. H. Kagermann, H. Johannes, H. Ariane, W. Wolfgang, *Recommendations for Implementing the Strategic Initiative INDUSTRIE 4.0: Securing the Future of German Manufacturing Industry*. Final Report of the Industrie 4.0 Working Group (Forschungsunion, Frankfurt, Germany, 2013)
8. AI for Good Global Summit, Geneva (2017) http://www.itu.int/en/ITU-T/AI/Pages/201706-default.aspx
9. United Nations: Looking to Future UN to Consider How Artificial Intelligence Could Help Achieve Economic Growth and Reduce Inequalities, http://www.un.org/sustainabledevelopment/blog/2017/10/looking-to-future-un-to-consider-how-artificial-intelligence-could-help-achieve-economic-growth-and-reduce-inequalities/2017

10. J. Ginsberg, M.H. Mohebbi, R.S. Patel, L. Brammer, M.S. Smolinski, L. Brilliant, Detecting influenza epidemics using search engine query data. Nature **457**(7232), 1012e4 (2009)
11. A. Mavragani, G. Ochoa, Google Trends in infodemiology and infoveillance: methodology framework. JMIR Public Health Surveill. **5**(2), e13439 (2019)
12. A.R. Daughton, M.J. Paul, Identifying protective health behaviors on Twitter: observational study of travel advisories and Zika virus. J. Med. Internet Res. **21**(5), e13090 (2019)
13. A. Signorini, A.M. Segre, P.M. Polgreen, The use of Twitter to track levels of disease activity and public concern in the US during the influenza A H1N1 pandemic. PLoS ONE **6**(5), e19467 (2011)
14. V. Gianfredi, N.L. Bragazzi, D. Nucci, M. Martini, R. Rosselli, L. Minelli, M. Moretti, Harnessing big data for communicable tropical and sub-tropical disorders: implications from a systematic review of the literature. Front. Public Health **6**, 90 (2018)
15. Air Transport Statistics 2018. International Air Transport Association (IATA), http://www.iata.org/services/statistics/air-transport-stats/Pages/index.aspx
16. N.L. Bragazzi, V. Gianfredi, M. Villarini, R. Rosselli, A. Nasr, A. Hussein, M. Martini, M. Behzadifar, Vaccines meet big data: state-of-the-art and future prospects. From the classical 3Is ("isolate-inactivate-inject") Vaccinology 1.0 to Vaccinology 3.0, vaccinomics and beyond: a historical overview. Front. Public Health **6**, 62 (2018)
17. J. Mossong, N. Hens, M. Jit, P. Beutels, K. Auranen, R. Mikolajczyk, M. Massari, S. Salmaso, G.S. Tomba, J. Wallinga, J. Heijne, M. Sadkowska-Todys, M. Rosinska, W. J. Edmunds, Social contacts and mixing patterns relevant to the spread of infectious diseases. PLoS Med. **5**(3), e74 (2008)
18. D. da Silva Motta, R. Badaro, A. Santos, F. Kirchner, Chapter 7: Use of artificial intelligence on the control of vector-borne diseases, in *Vectors and Vector-Borne Zoonotic Diseases*, ed. by S. Savic (IntechOpen, United Kingdom, 2018). ISBN 978-1-78985-293-6
19. C.S. Malley, J.C. Kuylenstierna, H.W. Vallack, D.K. Henze, H. Blencowe, M.R. Ashmore, Preterm birth associated with maternal fine particulate matter exposure: a global, regional and national assessment. Environ. Int. **101**, 173–182 (2017)
20. B. Wahl, A. Cossy-Gantner, S. Germann, N.R. Schwalbe, Artificial intelligence (AI) and global health: how can AI contribute to health in resource-poor settings? BMJ Glob. Health **3**, e000798 (2018)
21. J.P. Munoz, R. Boger, S. Dexter, J. Li, R. Low, Image recognition of disease-carrying insects: a system for combating infectious diseases using image classification techniques and citizen science, in *Proceedings of the 51st Hawaii International Conference on System Sciences* (HICSS, 2018), pp. 2835–2844. ISBN 978-0-9981331-1-9
22. G. Fleming, M. Mvander, G. McFerren, Fuzzy expert systems and GIS for cholera health risk prediction in southern Africa. Environ. Model. Softw. **22**, 442–448 (2007)
23. G. Eysenbach, Infodemiology and infoveillance: framework for an emerging set of public health informatics methods to analyze search, communication and publication behavior on the internet. J. Med. Internet Res. **11**(1), e11 (2009)
24. G. Eysenbach, Infodemiology and infoveillance: tracking online health information and cyber-behavior for public health. Am. J. Prev. Med. **40**(5), S154–S158 (2011)
25. H.T. Ho, T.M. Carvajal, J.R. Bautista, J.D.R. Capistrano, K.M. Viacrusis, L.F.T. Hernandez, K. Watanabe, Using Google Trends to examine the spatio-temporal incidence and behavioral patterns of dengue disease: a case study in metropolitan Manila, Philippines. Trop. Med. Infect. Dis. **3**, 118 (2018)
26. C. Alicino, N.L. Bragazzi, V. Faccio, D. Amicizia, D. Panatto, R. Gasparini, G. Icardi, A. Orsi, Assessing Ebola-related web search behavior: insights and implications from an analytical study of Google Trends-based query volumes. Infect. Dis. Poverty **4**, 54 (2015)
27. N. Mahroum, M. Adawi, K. Sharif, R. Waknin, H. Mahagna, B. Bisharat, M. Mahamid, A. Abu-Much, H. Amital, N.L. Bragazzi, A. Watad, Public reaction to Chikungunya outbreaks in Italy—insights from an extensive novel data streams-based structural equation modeling analysis. PLoS ONE **13**(5), e0197337 (2018)

28. O. Oluwagbemi, E. Adeoye, S. Fatumo, Building a computer-based expert system for malaria environmental diagnosis: an alternative malaria control strategy. Egypt. Comput. Sci. J. **33**(1), 55–69 (2009)
29. A. Sheikhtaheri, F. Sadoughi, Z.H. Dehaghi, Developing and using expert systems and neural networks in medicine: a review on benefits and challenges. J. Med. Syst. **38**, 110 (2014)
30. A. Caliskan, J.J. Bryson, A. Narayanan, Semantics derived automatically from language corpora contain human-like biases. Science **356**, 183–186 (2017)
31. J.L.K. Angwin, S. Mattu, L. Kirchner, *Machine Bias* (ProPublica, 2016)
32. R. Moss, A.E. Zarebski, S.J. Carlson, J.M. McCaw, Accounting for healthcare-seeking behaviors and testing practices in real-time influenza forecasts. Trop. Med. Infect. **4**(1), 12 (2019)
33. IEEE Symposium, *Algorithmic Transparency via Quantitative Input Influence: Theory and Experiments with Learning Systems. Security and Privacy (SP)* (IEEE, 2016)
34. A. Shaban-Nejad, M. Michalowski, D.L. Buckeridge, Health intelligence: how artificial intelligence transforms population and personalized health. NPJ Digit. Med. **1**, 53 (2018)
35. S. Feng, K.A. Grepin, R. Chunara, Tracking health seeking behavior during an Ebola outbreak via mobile phones and SMS. NPJ Digit. Med. **1**(1), 51 (2018)
36. K. Ganasegeran, S.A. Abdulrahman, Adopting m-Health in clinical practice: a boon or a bane?, in *Telemedicine Technologies*, ed. by H.D. Jude, V.E. Balas (Elsevier Academic Press, United States, 2019), pp. 31–41

Novel Non-contact Respiration Rate Detector for Analysis of Emotions

P. Grace Kanmani Prince, R. Immanuel Rajkumar and J. Premalatha

Abstract Emotions can be recognized by utilizing Physiological parameters such as pulse rate, respiration rate, measure of perspiration, conductance of skin and blood pressure. One of the strategies to study emotions is to analyze the variations in respiration rate with respect to change in emotions. It is noted that the respiration rate increases with increase in anxiety and slows down when the person is calm. An extensive review of how respiration is related to emotions is carried out in this work. A non contact respiration rate sensor is designed to obtain the respiration rate with much ease and accuracy when compared to other conventional respiration rate sensors. An algorithm is developed which maps the respiration rate and the emotion of an individual.

Keywords Respiration rate · Emotions · Non contact respiration detector · Symlet wavelet transform and IR sensor

1 Introduction

Human emotions and behavior patterns are put to rigorous study. Emotions are impressions that are felt by an individual and it relies upon numerous components like state of mind, encompassing conditions, happenings, inward body conditions and instincts. Emotions can likewise influence the body condition and could cause certain medical problems. Asthma gets triggered by strong negative or positive emotional changes [1]. Depression directly deals with emotions [2]. Stressful lifestyle with negative emotions can also cause heart disease and for patients with

P. G. K. Prince (✉) · R. Immanuel Rajkumar · J. Premalatha
Sathyabama Institute of Science and Technology, Chennai, India
e-mail: coggrace05@gmail.com

R. Immanuel Rajkumar
e-mail: imman047@gmail.com

J. Premalatha
e-mail: lathaaram@gmail.com

D. J. Hemanth (ed.), *Human Behaviour Analysis Using Intelligent Systems*,
Learning and Analytics in Intelligent Systems 6,
https://doi.org/10.1007/978-3-030-35139-7_8

cardiac problem, it could even worsen the health status [3]. Emotions also have an effect on ear problem which is called tinnitus. It is mainly caused by stressful feelings, anxious thoughts and depression which are all connected to the emotional centre of the brain [4, 5]. Gastric ulcers have also been related to emotions [6]. These are a few instances where the emotions influence the physiological and mental wellbeing of the human body system.

The emotion can not be assessed using a straight forward strategy. Thus change in physiological parameters identified with emotions are broken down into different parameters. One of the significant parameters considered is the heart rate in which the heart rate variability is read for different emotions [7]. Heart sounds [8] are placed into a test to identify if there are any progressions during changing emotions. The next parameter is blood pressure. It is presumed that blood pressure can be straightforwardly connected to emotion and emotional arousal is connected with the systolic and diastolic pressure in the human body [9]. Sweat is likewise a marker to recognize the physiological impact of emotion [10]. In view of the emotional changes, the perspiring shifts and palm-perspiring are also observed to show variations with change in emotion. It finds its application in detection of lie and assessment of stress levels. Likewise, variation of skin conductance has been identified with the investigation of emotions [11].

Different examinations have demonstrated that respiration rate is likewise modified by emotions. This chapter deals with a top to bottom investigation of how respiration rate can be utilized as a marker for analysis of emotions. A study of the fact of how the brain is responsible for the alteration of respiration during the surfacing of various emotions is done. A part of this chapter deals with sensors for estimating respiration rate. It introduces a novel non-contact breath rate sensor which accurately measures the respiration rate. The accessible information are mapped to relate how respiration changes for various types of emotions. The sensor can also be used to study the pattern of inhalation and exhalation for each emotion.

2 Anatomy and Physiology of Brain that Relates to Emotion and Respiration

2.1 Portion of the Brain Related to Emotions

The brain contains four lobes namely frontal, parietal, occipital and temporal which are responsible for the mental and physical activities of the human body. The temporal lobe is responsible for emotions. Beneath the temporal lobe and on either sides of the thalamus, which forms the mid part of the brain is located the limbic structure. This is called the emotional centre of the brain. It is responsible for behaviour patterns, long term memory, motivation, sense of smell etc. Thus the limbic system present in the brain is the most significant part which relates to emotions.

The Limbic system has a profound structure which comprises of temporal lobe and it is connected with the bottom edge of the frontal lobe. This system aids in amalgamating primary emotions and other mental capabilities. Hence it is also named as "Emotional nervous system". This part of the brain is what makes the activities pleasurable or memorable. Thus it is a link that connects the physical condition with the mental state of the human being. This is the reason for having high blood pressure due to high mental stress.

The limbic system has many parts namely amygdala, hippocampus, prefrontal cortex, hypothalamus, cingulate gyrus, and ventral tegmental area. The amygdale present in the limbic system is known as the emotional central processing portion of the brain. It can be found on either side of the temporal lobes. This is a significant region that aids in recognizing threats and prepares the mind for defensive mechanism. Studies have been carried out which proves that the amygdale initiates the change in certain body parameters corresponding to various emotions. It is connected to the autonomous nervous system (ANS). It includes the sympathetic and parasympathetic nervous system. ANS is responsible for voluntary and involuntary activities. Hence there is a change in the body parameters with respect to emotions. It has the ability to alter respiration rate, skin conductance and other physiological parameters as a response to emotions [12, 13]. Figure 1 shows the anatomy of the limbic system which is responsible for emotions [14].

The next part of the limbic system is hippocampus, which lies buried in the temporal lobe. It mainly deals with the formation of memories. It is also related with social emotions [15]. The hypothalamus initiates the emotions based on the visual information. For example, the fear that is caused by looking at horrific pictures will trigger the emotional changes [16]. Cingulate Gyrus deals with emotions which have a conscious experience [17]. The ventral tegmental area has a connection with the emotions related to love [18].

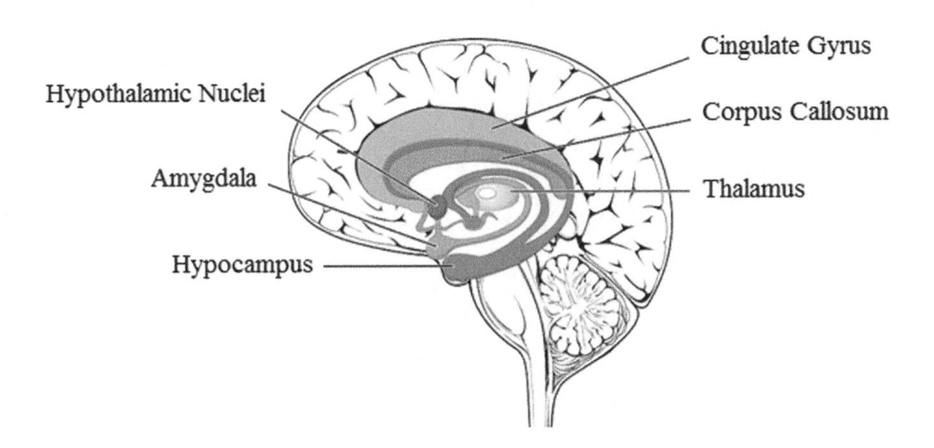

Fig. 1 Limbic system of the brain. *Source* Anatomy & Physiology, Connexions

2.2 Portion of the Brain Related to Respiration

The medulla oblongata and pons are the centres in the brain responsible for controlling physiological parameters that are essential for the existence of life. For example functioning of heart, breathing and swallowing are involuntary functions that are vital. It forms a portion of the brain stem which links the brain and the spinal cord. Figure 2 depicts the portion of the brain which is responsible for respiration [19].

Respiration occurs because of the working of the respiration focuses. There are three groups of neuron that are in charge of respiration. Two groups are found in the medulla and the third is found in the pons. The medulla includes Dorsal respiratory gathering and Ventral respiratory group. The dorsal respiratory group is responsible for emitting periodic bursts of the action potential which is responsible for inspiration. This part of the brain takes up the responsibility of maintaining the respiratory rhythm. The ventral respiratory group controls both inspiration and expiration. During strenuous exercises, the expiration will be intense. The pontine respiratory group is found in the pons and it has two segments pneumotaxic and apneustic regions. The Pontine respiratory group is responsible for the voluntary respirational activity of both inspiration and expiration [20].

The mechanism of respiration in a simplified form is given in Fig. 3. The brain stem where the pons and medulla are present, forms the network for respiration and they control the expiration and inspiration. The peripheral afferent information received from the receptors also influence breathing. The signal from the brain stem is given by the sympathetic nervous system and through the spinal cord, the signals are given to the respiratory muscles to contract and relax.

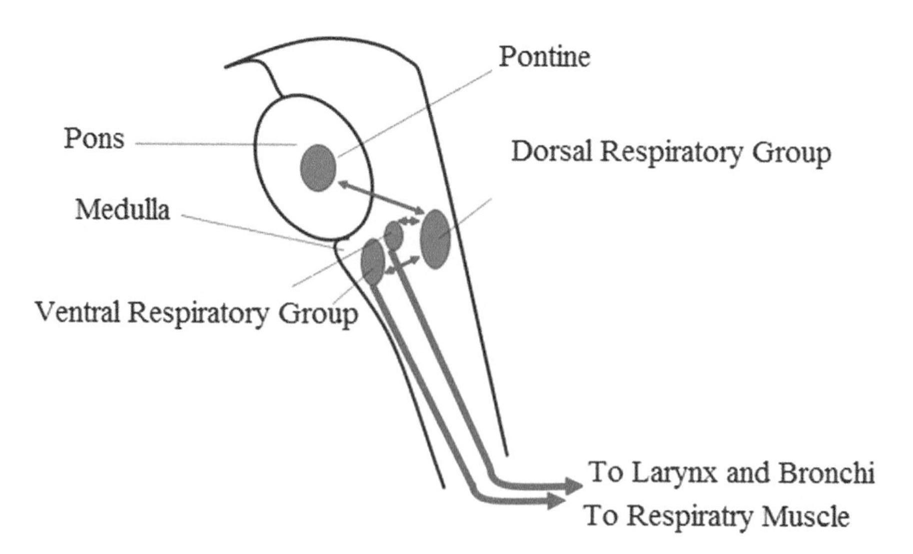

Fig. 2 Portion of the brain that is responsible for respiration. *Source* Neural Control of breathing

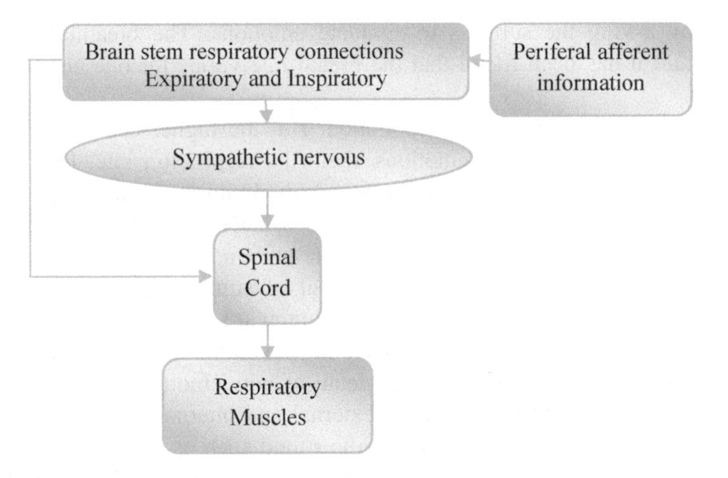

Fig. 3 Mechanism of respiration

Respiration is also influenced by the interaction with its connection to the environment. For example, to smell any object, the air has to be inspired. The inhalation takes a longer duration when the sense of smell is pleasant and for the unpleasant smell, the inhalation duration becomes too short when compared to the normal inhalation time.

2.3 *Emotion and Respiration*

The relation between emotion and respiration was studied as early as 1916 by Feleky [21]. He has experimented with the duration of inspiration and expiration. He studies have revealed that the duration of expiration is longer than the inspiration time. For certain emotions like happiness and sadness, the inspiration time was lesser than the expiration time. Feelings like disgust and anger had a longer duration of inspiration when compared to expiration. Respiration is normally channelized by the metabolic or balancing needs of the body but it is also influenced by the change in emotions. This indicates that there is a synergy between the lower brain stem and higher centres of the brain. The limbic system and cortex which is responsible for emotions also influence the rhythm of breathing or respiration [22].

Emotions are related to the central nervous system but they trigger the peripheral nervous system as well which leads to the change in the physiological parameters of the body. Hence Respiration is one parameter which can be used as a marker for panic attacks [23]. Both positive and negative emotions are related to the change in ANS activities like breathing, heart rate, etc. [24]. The study on emotions can also be done by asking the subjects to simulate emotions. The breathing pattern is observed and it has been found that the breath rate varies for one emotion to the other [25].

The metabolic rate influences respiration. But the higher centres of the brain which is also responsible for emotions alters the respiration rate. But when the carbon dioxide content increases in the blood, the brain stem takes over the complete control of respiration to maintain a proper balance of oxygen saturation in the blood by inspiring more air and expiring more carbon dioxide.

Emotions are formed due to some arousal of stimulus from the external or internal environment. For example, if any untoward incident takes place, it is natural to be emotionally sad and if something favorable happens, then happiness surfaces as the emotion. Figure 4 gives detail of how emotion is processed in the brain. Emotions are aroused by certain external and internal environment. Seeing something that is scary or hearing a horrific sound such as experiencing pain is a stimulus that causes a negative effects. Seeing a very beautiful flower or feeling a cool breeze causes positive emotions. These stimulus or arousal are given as input to the brain by the sense organs. These inputs are processed by the brain and the brain appraises the condition and sends the signal to the nervous system which causes a change in the physiological parameters such as respiration rate and heart rate. It also causes changes in the actions according to the situation like initiating self-defence mechanism or feeling relaxed, etc. Hearing a horrific sound or seeing any dangerous thing causes one's heart rate and respiration rate to increase and as a self-defence, the hands reach to ears to block the sound and the feet runs to get away from the danger. At the same time, the emotion centre of the brain is activated and causes a change in the mental state which is expressed as emotion. Unpleasant happenings cause fear, anxiety, sadness, frustration, etc. Pleasant happenings cause emotions such as happiness, calmness, laughter, etc. Thus changes in the physiological parameters such as heart rate, respiration rate, skin conductance and amount of sweat can aid in quantifying emotions. Of these, respiration rate is more is easily accessible and could give a meaningful analysis of emotions.

3 Respiration Rate Measurement

3.1 Respiration Rate Measuring Devices

The respiration rate can be measured by using various devices. Strain gauge sensor is used in a belt type setup which is wrapped around the chest. The resistance varies due to the movement of the diaphragm during inhalation and exhalation. Due to the variation in the resistance, the voltage changes and the occurrence of peak per minute is recorded which gives the respiration rate.

Thermistor is used to detect the change in temperature due to expiration and inspiration [26]. The thermistor is placed in the nasal cavity so that the temperature

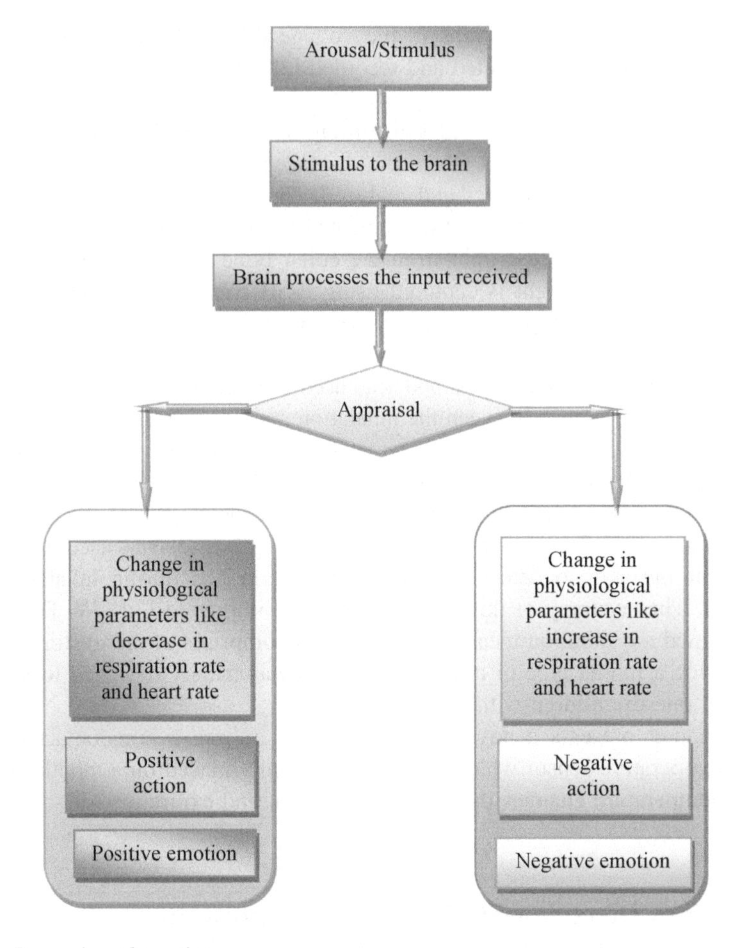

Fig. 4 Processing of emotion

changes can be noted with accuracy. But it will not be much comfortable to place the sensor inside the nostril. These are the two types of sensors that are most commonly used for respiration measurement. The above two sensors are contact type since they are placed on the body itself. This might cause discomfort to the subjects. Hence the non-contact type of respiration rate measurements was studied.

Another method that is used in detecting the sound that comes while inhaling and exhaling. A miniaturized microphone is placed near the nostrils and the output voltage from the microphone is analyzed and the peaks per minute will be counted.

The flow rate of the inhaled air and exhaled air is measured to study the pattern of respiration [27]. In this method, a miniaturized pitot tube arrangement is placed in the nasal cavity with the sealed nostrils. The nasal pressure is monitored in this method. Preparation of handling the sensor is a tedious task but accuracy will be high. A single-lead ECG system placed on the chest can detect the respiration rate but it not accurate. The variations in the voltage due to the chest measurement during respiration can be detected and recorded [28]. But the results are not much accurate.

One of the methods employ video recording using RGB recording. The image processing techniques are used where the displacement of the chest due to inspiration and expiration are recorded and analyzed. The Principal Component Analysis is employed for analysing the chest movements [29]. Another method is to measure the intensity of the reflected light from the collar bone which moves during respiration. The reflected light is captured using RGB camera. The image that was processed in the frequency domain gave more accurate results when compared to the time-domain method [30].

The infrared thermographic method was used to find the respiration rate where the temperature difference between the inhalation and exhalation is very minimum. The method was used in infants. This method proved to be more accurate. The disadvantage is that the image acquisition and processing time were too large and another inconvenience was that the instrument was too large and occupied considerable space [31]. In another method, a temperature sensor was used which was placed near the skin to detect the temperature variations during inhalation and exhalation. The accuracy of this method had to be improved [32]. W. Daw et al. have designed a respiration measuring device that comprises of thermistor which is placed inside a funnel type of arrangement. This was mainly designed to note the respiration rate of children with much ease [33].

This chapter presents a novel approach which utilizes a non-contact type of infrared temperature sensor which can be conveniently placed on a mouth mask to measure temperature changes during the inspiration and expiration.

4 Emotion Detection Studies Using Respiration Measurement and Computer Interface

A study was conducted by de Melo et al. on virtual humans as to how the respiration pattern varies with different emotions [34]. This study includes exhaustive research on the breathing patterns for fourteen different emotions on a computer model and subjects were asked to assess the model to improve it for further studies.

An extensive study of how Human-computer interface can be used to study emotions for tapping the benefits of technology was studied by Voeffray [35]. The suggestion has been made that visual or auditory or physiological means can be used to automate the detection of emotions. But the disadvantage is that the subject has to be kept wired to access the physiological data which might cause some inconvenience to the subject.

Another analysis of connecting respiration and emotions were done using two approaches [36]. In the first approach, the participants in the study were asked to generate various types of emotions and it confirmed that it was comparable with psychophysiological trials. The second approach was by analyzing the emotions and respiration when the participants were shown a cover story and their emotional changes were noted.

In an exhaustive research, Physiological signals such as ECG, skin temperature, skin conductance and respiration rate was employed to study the variations of physiological parameters with respect change in emotion [37, 38]. To induce emotion a film clipping was shown and the emotions were studied. The canonical correlation was used to classify the data set and this method obtained an accuracy of 85.3%.

The airflow of exhaled and inhaled air measurement is taken to study the flow pattern of air through the nostrils for various types of emotions [39]. It has been noted the inspiration and expiration of air are more and the process happens more slowly when the subject is relaxed but it varies faster for negative emotions. For anger, the inspiration and expiration of air is more and rate is faster but for fear, inspired and expired air is comparatively lesser but the rate is faster than the normal.

Thus these studies prove that recognition of emotions based on the physiological parameters such as respiration rate along with Human Machine interface will aid in numerous applications.

5 Respiration Rate Measurement Using Non-contact Temperature Sensor

The hardware setup consists of a non-contact infrared temperature sensor MLX90614DAA. This sensor has an accuracy of ±0.5 °C which can be used for medical purposes. The temperature range of the sensor for an object is 32–42° to suit the medical applications. The ambient temperature range for which it works is 10–40 °C. It is an I2C device, hence retrieval of signals is made easier.

This miniaturized device consists of mainly two parts. One is the thermopile which detects the infrared radiations emitted by the object and the other is the signal conditioning circuit which converts the output voltage of the thermopile into a voltage range that can be calibrated in terms of temperature. This sensor follows the law of Stefan-Boltzmann. It indicates that all objects that have a temperature above absolute zero will emit infrared radiation which is proportional to the temperature of the object. This sensor captures this infrared radiation that falls along the field of view of the sensor. The thermopile converts these radiations into an electrical voltage which is proportional to the temperature. The voltage from the thermopile is given to a 17 bit Analog to Digital converter and is then given to a signal conditioning circuit. It is then finally given for processing to the microcontroller.

This sensor can read the ambient temperature as well as the object temperature, the ambient temperature can also be used for calibration. Figure 5 shows the noncontact type of temperature measuring device which is MLX90614DAA which signifies that it is calibrated for medical usage.

Fig. 5 MLX90614DAA sensor used for measuring temperature

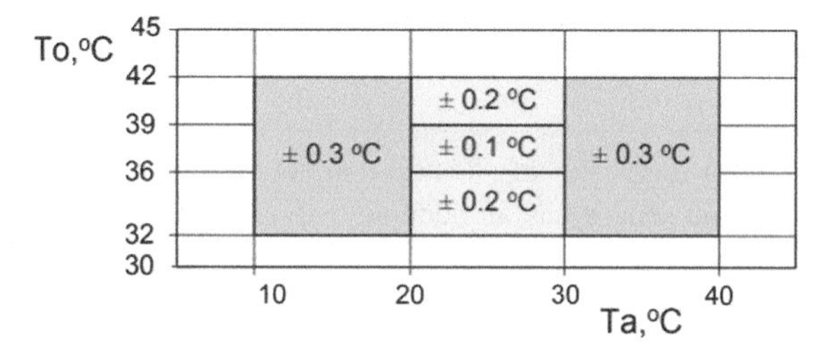

Fig. 6 Accuracy table for MLX90614DAA for ambient temperature and object temperature. *Source* SEN-09570-datasheet-3901090614M005.pdf

The signal is acquired using the microcontroller ATMEGA328P. The output of the sensor is read which directly gives the accurate temperature of the object which is in proximity to the sensor. The output of the sensor is the voltage signal which is proportional to the temperature and the output voltage is calibrated to give the exact temperature. The signal acquisition is done and the data are collected through the processing software. The respiration rate is displayed on the computer screen. Figure 6 shows the accuracy of MLX90614DAA sensor for object temperature To and room or ambient temperature Ta. It clearly shows that the accuracy for temperature ranges of 32 and 42 °C of the object has an accuracy of ±0.3 when the ambient temperature is from 10 to 20 °C and 30 to 40 °C. The accuracy for temperature range becomes lesser for 20–30 °C. It can vary from ±0.1 and ±0.2.

6 Measurement of Respiration Rate

The sensor is placed in the mouth mask in proximity to the nostrils so that the temperature variations in the space near the nostrils while inhaling and exhaling could be noted. Generally, the temperature of the exhaled air is slightly higher than the ambient temperature or the body temperature at the surface of the skin. While exhaling the voltage gradually increases and while inhaling the voltage gradually reduces. This cycle repeats itself for each respiratory cycle. Hence the number of rises in the output voltage per minute provides the respiration rate or the breath rate. The algorithm for calculating the respiration rate is given below.

Algorithm

1. As the data streams into the controller, 50 number of consecutive data are stored in an array.
2. The maximum and minimum values in the array are noted.
3. A threshold level is fixed below the highest value in the array.
4. When the data the point crosses the threshold level and when it falls below the threshold, the count is updated. Positive and the negative slope are taken into account for the count to be updated.
5. The count is taken for 15 seconds and then converted into minute count.

The major advantage of this algorithm is that it is patient-specific since it takes the peak value for each subject and then calculates the respiration rate. The time taken to measure the respiration rate is very less which is around 20 s.

The output is sinusoidal in nature since the rise in temperature during exhalation is gradual and it reduces gradually during inhalation. It is noted that graph is taken when the person is relaxed and happy, the slope of the signal is less when compared to signals which were acquired when the person is anxious with fear or any other negative emotions.

7 Variation of Respiration Pattern with Respect to Emotions

The pattern of the signal varies from person to person and from emotion to emotion. Usually, it was observed that the waveform or signal follows a sinusoidal pattern. As discussed earlier respiration is altered with respect to emotions. Ten normal subjects were taken for studying the effects of emotion on the rate of respiration. Two types of observations were made. Firstly the subjects were asked to simulate emotions. When they had done so, their expression also changes and the respiration sensor present in the mouth mask was worn by the subject. Thus the respiration rate for different emotions were acquired. It is also noted that the values of respiration rate when the emotions are simulated are higher when compared to the respiration

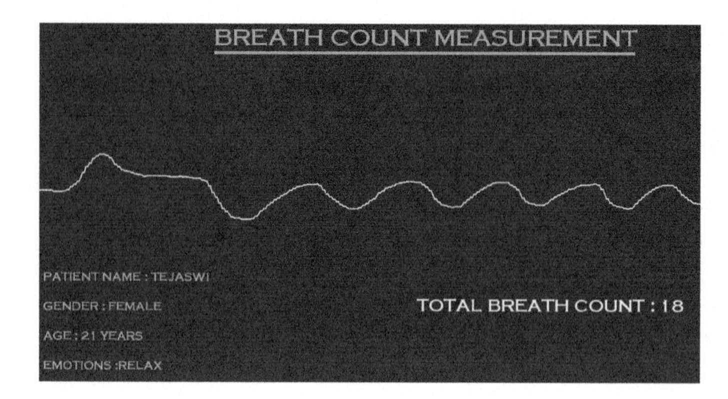

Fig. 7 Respiration pattern when the subject is happy and relaxed

rate that would occur under unsimulated real circumstances. Depending on the fierceness of the emotion, the respiration rate could vary. Figure 7 shows the recording of the respiration when the person is relaxed. That is, the emotion can be taken as happy. This waveform clearly shows that it is a slowly varying sinusoidal signal with comparatively low amplitude and low frequency. The slope of the signal is much lesser which indicates that the temperature variation during inhalation and exhalation is quite lesser. The respiration rate is around 18.

Figure 8 displays the pattern when the subject was asked to simulate anger. It is clearly seen that during this time the amplitude and frequency are higher when the subject relaxed and normal. It was also observed that in some cases, during the time of anger or for an emotion like fear the amplitude was lesser and the frequency remained higher. The slope of the waveform is much higher when compared to signals acquired when the subject is relaxed.

The next sample of respiration rate measurement shows the variation for the same subject when the emotion is happy and when the subject is in anger. This is shown in Fig. 9. In this case during the state of anger, the inhalation and exhalation

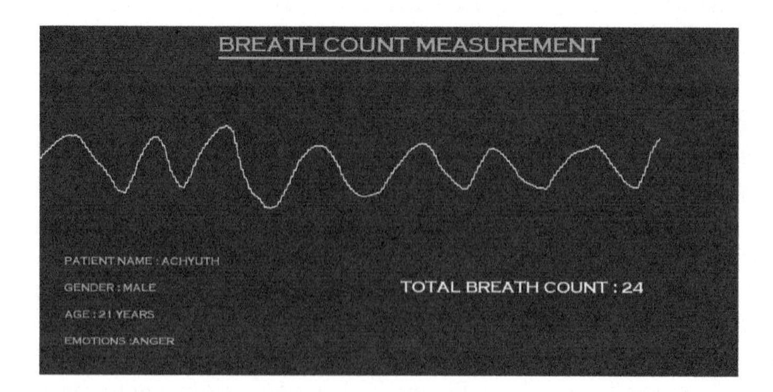

Fig. 8 Respiration pattern when the subject is angry

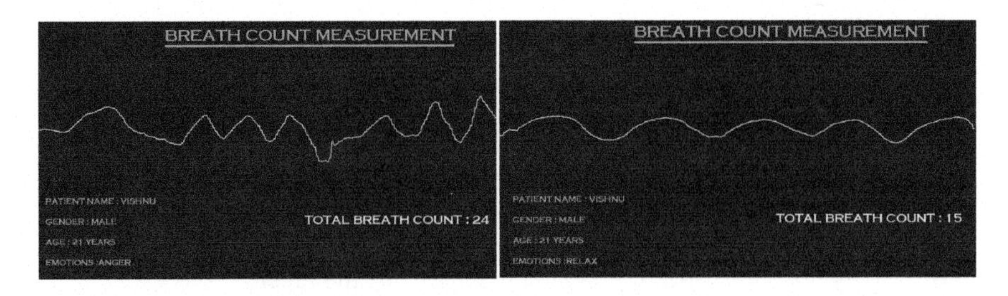

Fig. 9 Respiration pattern of the same subject for anger and happy

cycle could get disturbed where as it is uniform when the subject is relaxed or in a pleasant mood which reflects the positive emotions.

The rate of respiration is more when the subject is anxious, in fear or anger. The rate of respiration is lesser in a relaxed state. The amplitude of the signal during anxiety is lesser and during a relaxed mood, it is higher since more amount of air is inhaled and exhaled.

8 Application of Wavelet Transform for Differentiating Positive and Negative Emotions

The signal is acquired from the sensor and the statistical features are extracted to see is they would aid in the classification of emotions.

Then the wavelet transform is taken using Symlet mother wavelet. The approximations and details are obtained. The statistical features of the details are retrieved and studied for positive and negative emotions. The level which gives the best variation is chosen. Figure 10 details about the steps that are followed for choosing useful features from the wavelet transformed signal.

The temperature variation during respiration is noted using a non-contact temperature sensor MLX90614 which is placed in the vicinity of the nostrils. It produces a voltage that is proportional to the temperature and the output is calibrated in terms of temperature. These signals that are acquired from the sensor is analyzed for positive and negative emotions. Positive emotion is when the subject is calm and relaxed. Happiness, joy, calmness, etc. are termed as positive emotions. When the subject is angry, anxious or perturbed, it is termed as negative emotions.

As said earlier, there is a shallow and fast response for negative emotion and deeper and slower ones for positive emotions. The subjects age group ranging from 20 to 40 were asked to volunteer for the study and the emotions were simulated by the subjects. Eight statistical parameters were taken from the acquired signal and were analyzed whether it could be used to differentiate between positive and negative emotions. The statistical parameters considered are mean, median, minimum value and maximum value, range, standard deviation, median absolute, mean

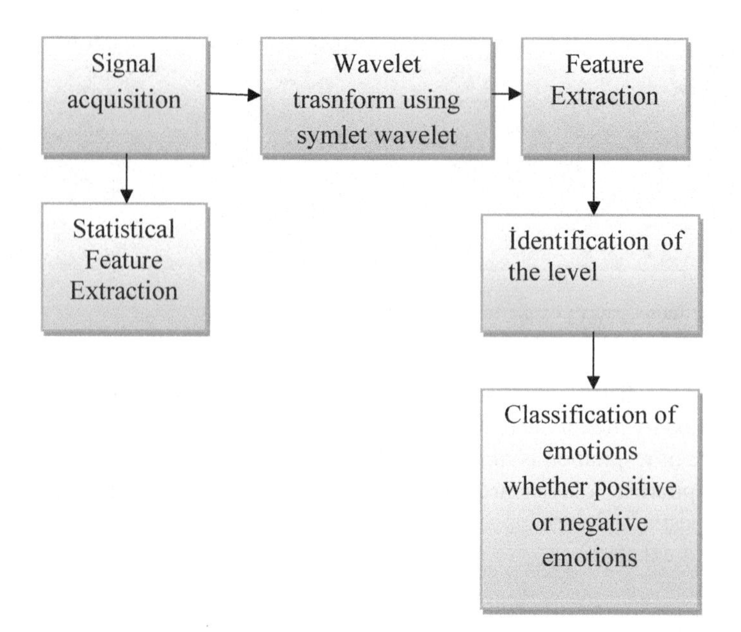

Fig. 10 Application of wavelet transform for differentiating positive and negative emotions

absolute deviation, L1 and L2 norm and max norm. Tables 1 and 2 shows the sample of values of mean, median, maximum, minimum and range of the signal are presented values of the signal that were taken when the subject was asked to relax and be happy.

Table 1 Mean, median, max, min and range (happy and relaxed)

Subject	Mean	Median	Max	Min	Range
S1	34.14	34.17	35.05	32.97	2.08
S2	33.89	33.91	34.97	33.05	1.92
S3	34.2	34.26	36.47	31.69	4.78

Table 2 Sandard deviation, median absolute, mean absolute deviation, L1 and L2 norm and max norm (happy and relaxed)

Subject	Std dev	Med abs dev	Mean abs dev	L1 norm	L2 norm	Max norm
S1	0.5715	0.52	0.5114	9559	571.3	35.05
S2	0.5097	0.42	0.4404	1.08E+04	605.4	34.97
S3	1.4	1.37	1.275	1.37E+04	684.6	36.47

Table 3 Mean, median, max, min and range (sad and anxious)

Subject	Mean	Median	Maximum	Min	Range
S1	34.47	34.33	36.85	31.75	5.1
S2	34.78	34.77	35.71	34.15	1.56
S3	34.08	34.13	35.05	32.77	2.28

Table 4 Sandard deviation, median absolute, mean absolute deviation, L1 and L2 norm and max norm (sad and anxious)

Subject	Std dev	Med abs dev	Mean abs dev	L1 norm	L2 norm	Max norm
S1	1.534	1.44	1.391	1.38E+04	690	36.85
S2	0.3496	0.24	0.2844	1.05E+04	603.5	35.71
S3	0.6121	0.54	0.5408	1.06E+04	600.1	35.05

The negative emotions such as being sad or with fear and anxiety, hence the above said statistical parameters were taken for negative emotions. Tables 3 and 4 provide the statistical parameters when the subject was asked to simulate sadness.

From the values given in the tables, the values of positive and negative emotion overlap each other hence it can not be effectively used in classification. Hence wavelet transforms is applied to the sensor signal.

The Symlet wavelet is chosen as the mother wavelet. Figure 11 shows the mother wavelet Symlet.

The Symlet wavelet 2 is applied to the signal and it is decomposed into 5 levels. The approximation and the details are extracted. The statistical parameters of level 5 are taken into consideration. The statistical parameters of the original system are not suited for the purpose of classification since all the values overlap. But the statistical values of the wavelet coefficient of detail level 5 gives promising results and this can be used for classification purpose. Figure 12 shows the wavelet transform of a subject for a signal acquired when the subject was relaxed and happy. Figure 13 shows the wavelet transform of a signal when the subject was asked to simulate an emotion of anger and sadness.

Figure 14 displays the colour coefficient graph of the subject in a relaxed state. The blue colour represents lower frequency coefficients and as the shade moves towards dark pink it indicates high-frequency components. Figure 15 shows the colour coefficient graph of the signal which was acquired when the subject had simulated a negative emotion like anxiousness due to fear, sadness, etc. It can be clearly inferred from the graph that emotion which is in the relaxed state has lesser frequency components when compared to the color coefficient graph of perturbed state. Both these signals were acquired from the same subject.

The Symlet wavelet is almost symmetrical wavelets which were developed by Daubechies who altered the Daubechies family of wavelets.

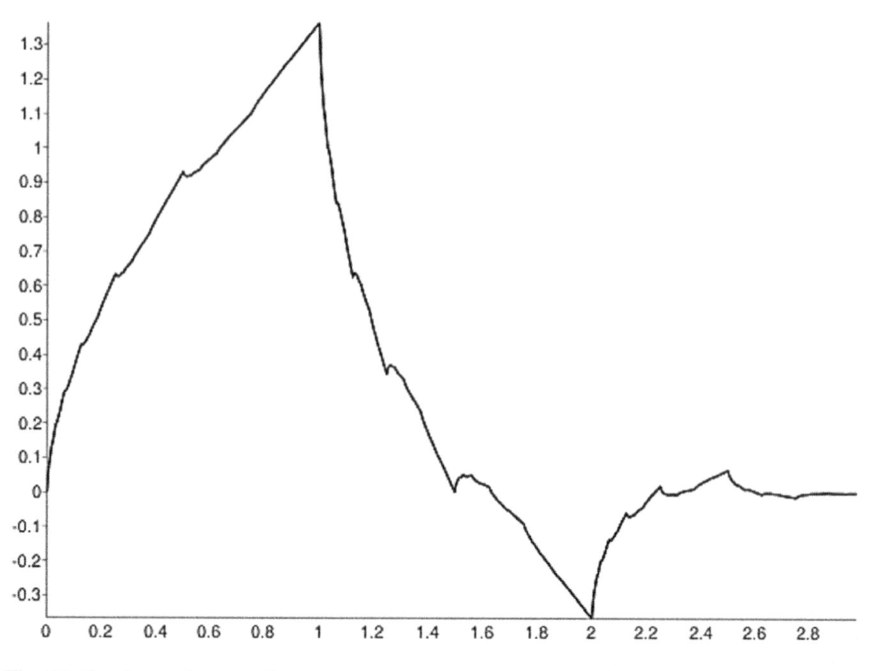

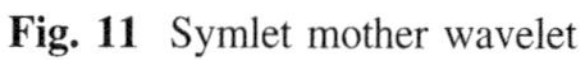
Fig. 11 Symlet mother wavelet

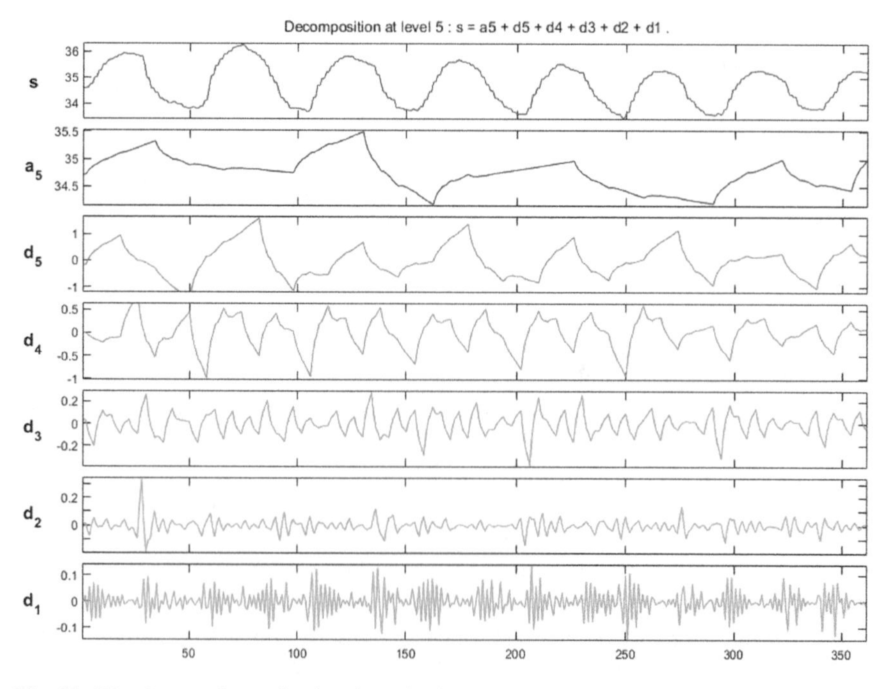

Fig. 12 Wavelet transform of a signal acquired when the subject is relaxed or happy

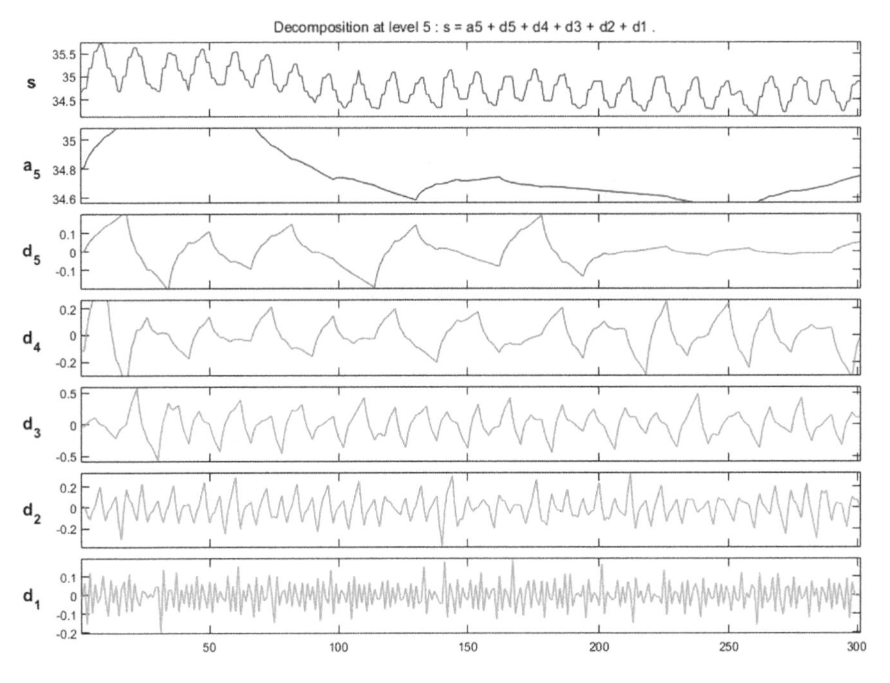

Fig. 13 Wavelet transform of a signal acquired when the subject is angry or sad

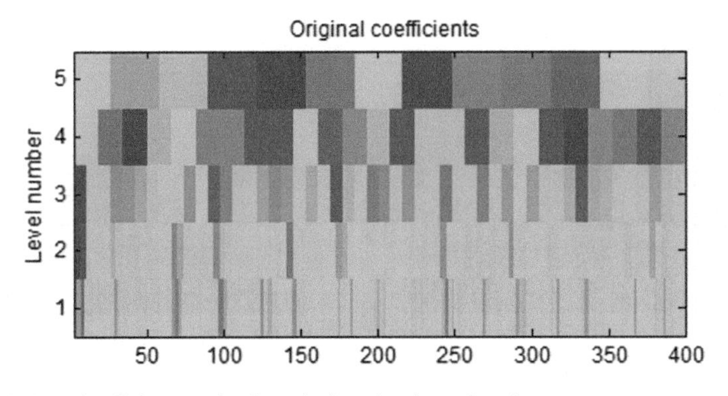

Fig. 14 Colour Coefficient graph of respiration signal at relaxed state

Figures 16 and 17 shows the level 5 of the detail of the wavelet transform with its statistical parameter for both positive and negative expressions. The statistical values of this level provide useful information for classifying signals based on emotions.

It is clearly seen that level 5 possesses information about the variation in the frequency of the signal. Tables 5 and 6 gives the statistical parameters for level 5 of

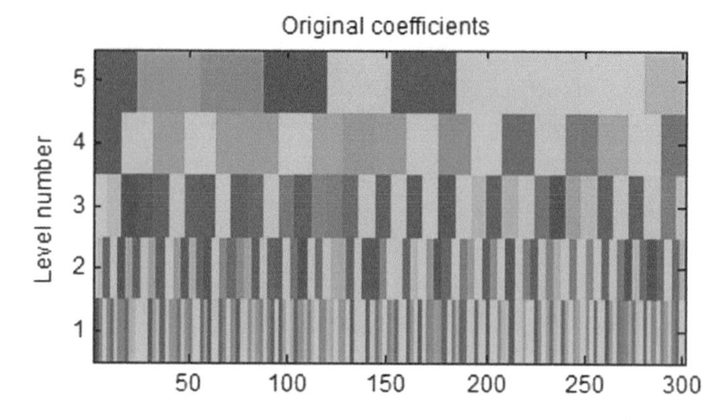

Fig. 15 Colour coefficient graph of respiration signal at anxious state

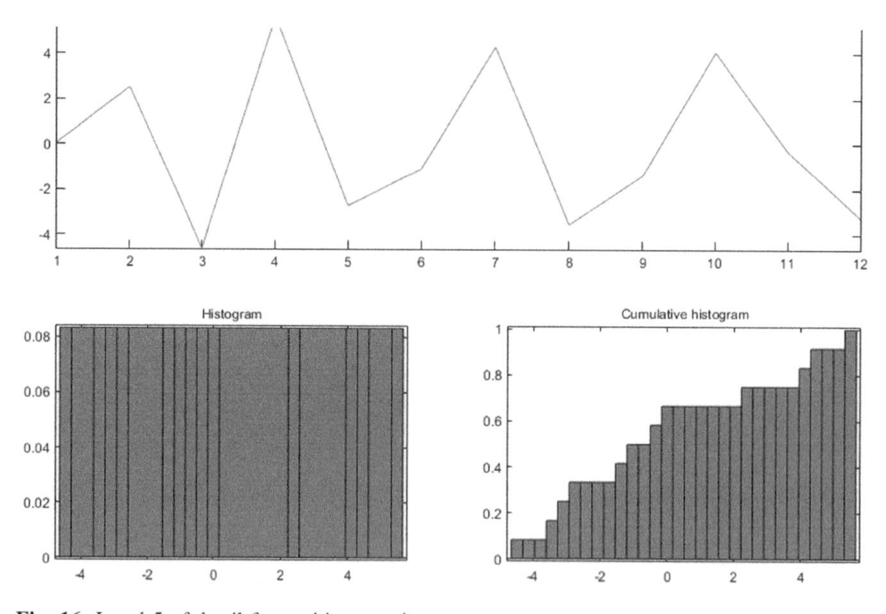

Fig. 16 Level 5 of detail for positive emotion

detail of wavelet transform under happy and relaxed condition. And Tables 7 and 8 gives the statistical parameters for level 5 of detail of wavelet transform under the sad state.

From Tables 5, 6, 7 and 8, it is clearly seen that the statistical values of level 5 provide a marked difference between the two emotional states. The mean value for positive emotion is lesser when compared to the negative emotions. The median between the emotional states overlap each other and hence cannot be used for classification. The rest of the parameter values are higher for positive state and

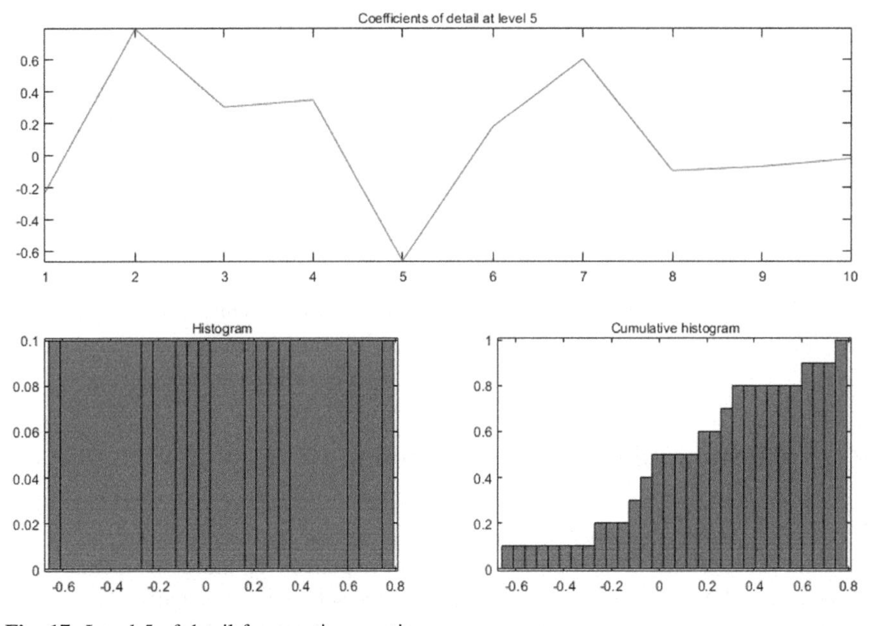

Fig. 17 Level 5 of detail for negative emotion

Table 5 Mean, median, max, min and range for level 5 of detail of wavelet transform (happy and relaxed)

Subject	Mean	Median	Max	Min	Range
S1	0.06138	−0.1273	3.584	−2.736	6.322
S2	−0.02019	−0.6895	5.666	−4.634	10.3
S3	0.0936	−0.21	7.075	−5.56	12.64

Table 6 Sandard deviation, median absolute, mean absolute deviation, L1 and L2 norm and max norm for level 5 of detail of wavelet transform (happy and relaxed)

Subject	Std dev	Med abs dev	Mean abs dev	L1 norm	L2 norm	Max norm
S1	2.299	2.183	1.833	16.31	6.51E+00	3.584
S2	3.423	2.706	2.794	33.56	11.35	5.666
S3	4.76	2.364	3.45	17.16	9.522	7.075

Table 7 Mean, median, max, min and range for level 5 of detail of wavelet transform (sad and anxious)

Subject	Mean	Median	Max	Min	Range
S1	0.2856	0.2112	0.8275	−0.2349	1.062
S2	0.1167	0.061	0.7932	−0.6599	1.453
S3	0.1044	−0.046	1.084	−1.406	2.49

Table 8 Sandard deviation, median absolute, mean absolute deviation, L1 and L2 norm and max norm for level 5 of detail of wavelet transform (sad and anxious)

Subject	Std dev	Med abs dev	Mean abs dev	L1 norm	L2 norm	Max norm
S1	0.4054	0.3257	0.3163	2.183	1.145	0.8275
S2	0.4231	0.2469	0.3306	3.606	1.322	0.7922
S3	0.8318	0.6776	0.7042	9.05	2.906	1.406

lesser for negative emotional state. Hence these parameters can be used as markers for classification of emotional states based on the temperature sensor signal.

9 Application of Emotion Recognition

The studies on emotion recognition are done for various purposes. It is applied in the medical field for analyzing the psychology of the patient. To know the mental state of persons who will not be able to express themselves like mentally challenged, neonatal, autistic children, etc. It can also be used to monitor the mental and physical state of old people and it also can aid in assessing the treatment given to patients are proving to be beneficial or not. It can be used as a marker for lie detection.

The emotion studies not only aid in the field of medicine but can also be used in the fields of Entertainment, and study the behavior pattern while using gaming applications, etc. The study can aid in identifying fatigue and give an alarm to the subject while doing a strenuous job and to send a warning to persons who are driving when they are stressed out. In call-centres emotion recognition is used to assess the stress levels of the speaker. In rare cases emotions can also be used as a biometric.

10 Conclusion

This chapter deals with an extensive study of anatomy and physiology of how respiration and emotions are linked. This study also introduces a novel approach to detect emotions using non-contact temperature measuring device which aids in the study of variation in the temperature during inhalation and exhalation for different emotions. The acquired signals were put to test by obtaining the statistical parameters. But the values of the statistical parameter overlapped which makes it tedious to classify the positive and negative emotions. Hence wavelet transform was used and was found that level 5 of detail provided useful inputs. The statistical parameters thus obtained from level 5 of detail of the wavelet transform provided a variation in the values of the statistical parameters. Hence using these techniques it would be easier to classify the signals using clustering techniques.

The future scope of this work is to collect more number of samples and acquire data for various types of emotions. The retrieved data can be used to classify the data using supervised or unsupervised learning methods. Automated recognition of emotions with high accuracy can be designed that would aid in the study of various Psychophysiological purposes.

References

1. K. Vazquez, J. Sandler, A. Interian, J.M. Feldman, Emotionally triggered asthma and its relationship to panic disorder, ataques de nervios, and asthma-related death of a loved one in Latino adults. J. Psychosom. Res. **93**, 76–82 (2017)
2. A. Llorca, E. Malonda, P. Samper, The role of emotions in depression and aggression. Med. Oral Patol. Oral Cir. Bucal **21**(5), e559–e564 (2016)
3. M. Bahremand, M. Alikhani, A. Zakiei, P. Janjani, A. Aghei, Emotion risk-factor in patients with cardiac diseases: the role of cognitive emotion regulation strategies, positive affect and negative affect (a case-control study). Glob. J. Health Sci. **8**(1), 173–179 (2015). https://doi.org/10.5539/gjhs.v8n1p173
4. University of Illinois at Urbana-Champaign, People with tinnitus process emotions differently from their peers, researchers report. ScienceDaily (2014 June 25). Retrieved 1 July 2019 from www.sciencedaily.com/releases/2014/06/140625184901.htm
5. K.C. Horner, The emotional ear in stress. Neurosci. Biobehav. Rev. **27**(5), 437–446 (2003)
6. M.H. Alp, J.H. Court, A.K. Grant, Personality pattern and emotional stress in the genesis of gastric ulcer. Gut **11**(9), 773–777 (1970). https://doi.org/10.1136/gut.11.9.773
7. S.-N. Yu, S.-F. Chen, Emotion state identification based on heart rate variability and genetic algorithm, in *2015 37th Annual International Conference of the IEEE Engineering in Medicine and Biology Society (EMBC)*, Aug 2015, pp. 538–541. https://doi.org/10.1109/embc.2015.7318418
8. C. Xiefeng, Y. Wang, S. Dai, P. Zhao, Q. Liu, Heart sound signals can be used for emotion recognition. Sci. Rep. **9**(1), 6486 (2019)
9. G.D. James, L.S. Yee, G.A. Harshfield, S.G. Blank, T.G. Pickering, The influence of happiness, anger, and anxiety on the blood pressure of borderline hypertensives. Psychosom. Med. **48**(7), 502–508 (1986)
10. Asahina, M., The neural mechanism of emotional sweating. Brain Nerve **68**(8), 883–892 (2016). https://doi.org/10.11477/mf.1416200526
11. E. Gatti, E. Calzolari, E. Maggioni, M. Obrist, Emotional ratings and skin conductance response to visual, auditory and haptic stimuli. Sci. Data **5**, Article number 180120 (2018)
12. Ueyama, T., Emotion, amygdala, and autonomic nervous system. Brain Nerve **64**(10), 1113–1119 (2012)
13. R.W. Levenson, The autonomic nervous system and emotion. Emot. Rev. **6**(2), 100–112 (2014)
14. Anatomy & Physiology, Connexions. Website: http://cnx.org/content/col11496/1.6/, 19 June 2013
15. M.H. Immordino-Yang, V. Singh, Hippocampal contributions to the processing of social emotions. Hum. Brain Mapp. **34**(4), 945–955 (2013). https://doi.org/10.1002/hbm.21485
16. K. Hirayama, Thalamus and emotion. Brain Nerve **67**(12), 1499–1508 (2015). https://doi.org/10.11477/mf.1416200328
17. H.G. Engen, M.C. Anderson, Memory control: a fundamental mechanism for emotion regulation. Trends Cogn. Sci. **22**(11), 982–995 (2018)
18. S. Edwords, *Love and the Brain. On the Brain* (The Harvard Mahoney Neuroscience Institute, 2019)

19. R. Carpenter, Neural Control of Brain. https://slideplayer.com/slide/5692388/
20. T.E. Dick, M. Dutschmann, K.F. Morris, Pontine control of respiration, in *Encyclopedia of Neuroscience*, ed. by M.D. Binder, N. Hirokawa, U. Windhorst (Springer, Berlin, Heidelberg, 2009)
21. A. Feleky, The influence of the emotions on respiration. J. Exp. Psychol. **1**(3), 218–241 (1916)
22. I. Homma, Y. Masaoka, Breathing rhythms and emotions. Exp. Physiol. **93**(9), 1011–1021 (2008)
23. A.A. Sarkar, Functional correlation between breathing and emotional states. MOJ Anat. Physiol. **3**(5), 157–158 (2017). https://doi.org/10.15406/mojap.2017.03.00108
24. M.N. Shiota, S.L. Neufeld, W.H. Yeung, S.E. Moser, E.F. Perea, Feeling good: autonomic nervous system responding in five positive emotions. Emotion **11**(6), 1368–1378 (2011). https://doi.org/10.1037/a0024278
25. P. Philippot, G. Chapelle, S. Blairy, Respiratory feedback in the generation of emotion. Cogn. Emot. 605–627 (2010)
26. B.B. Yearmwar, M. Mhetre, Respiration rate measurement: a different approach. Int. J. Pure Appl. Math. **118**(24) (2018)
27. G. Rosengarten, B. Ahmed, C. Marselli, A novel method to measure respiratory airflow rate based on nasal pressure. IFAC Proc. Vol. **36**(15), 415–419 (2003)
28. S.X.Z. Ding, W. Chen, D. Wei, Derivation of respiratory signal from single-channel ECGs based on source statistics. Int. J. Bioelectromagn. **6**, 43–49 (2004)
29. C. Wiede, J. Richter, M. Manuel, G. Hirtz, Remote respiration rate determination in video data—vital parameter extraction based on optical flow and principal component analysis, in *VISAPP 2017—International Conference on Computer Vision Theory and Applications* (2017)
30. C. Massaroni, D.S. Lopes, D. Lo Presti, E. Schena, S. Silvestri, Contactless monitoring of breathing patterns and respiratory rate at the pit of the neck: a single camera approach. J. Sens. Article ID 4567213 (2018)
31. C.H. Hsu, J.C. Chow, Design and clinic monitoring of a newly developed non-attached infant apnea monitor. Biomed. Eng. **17**, 126–133 (2005)
32. A.K. Abbas et al., Neonatal non-contact respiratory monitoring based on real-time infrared thermography. Biomed. Eng. Online **10**, 93 (2011). https://doi.org/10.1186/1475-925X-10-93
33. W. Daw, R. Kingshott, R. Saatchi, D. Burke, A. Holloway, J. Travis, R. Evans, A. Jones, B. Hughes, H. Elphick, Medical devices for measuring respiratory rate in children: a review. J. Adv. Biomed. Eng. Technol. **3**, 21–27 (2016). ISSN: 2409-3394/16
34. C.M. de Melo, P. Kenny, J. Gratch, Real-time expression of affect through respiration. Comput. Anim. Virtual Worlds **21**, 225–234 (2010)
35. C. Voeffray, Emotion-sensitive human-computer interaction (HCI): state of the art. Seminar paper, University of Fribourg, Fribourg Switzerland
36. P. Philippot, G. Chapelle, S. Blairy, Respiratory feedback in the generation of emotion. Cogn. Emot. **16**, 605–627 (2010). https://doi.org/10.1080/02699930143000392
37. L. Li, J.-H. Chen, Emotion recognition using physiological signals, in *International Conference on Artificial Reality and Telexistence, ICAT 2006: Advances in Artificial Reality and Tele-Existence*, pp. 437–446 (2006)
38. S. Balters, M. Steinert, Capturing emotion reactivity through physiology measurement as a foundation for affective engineering in engineering design science and engineering practices. J. Intell. Manuf. **28**(7), 1585–1607 (2017)
39. R.A. Hameed, M.K. Sabir, M.A. Fadhel, O. Al-Shamma, Human emotion classification based on respiration signal, in *ICICT'19. Proceedings of the International Conference on Information and Communication Technology* (2019)

Human Emotion Classification Using EEG Signals by Multivariate SynchroSqueezing Transform

S. T. Veena and M. N. Sumaiya

Abstract Communication between human and computer is highly demandable in different applications. Possessing of cognitive skills of a human for human-computer interaction (HCI) is a non-trivial task. Classification of human emotion involving artificial intelligence is an emerging field. HCI finds applications in smart home, industry, personal health and for different purposes. One such HCI task is an automated human emotion recognition system. In this paper, new human emotion classification method based on Electroencephalogram (EEG) signal is proposed that leverages singular value decomposition (SVD) and Multivariate SynchroSqueezing Transform (MSST). Highly random, non-stationary EEG signal illustrates the electrical activity of the brain and contains useful information. To extract the hidden information/features, Multivariate SynchroSqueezing Transform is exploited. It is an adaptive, invertible transform to improve the quality of the time-frequency representation (TFR) by considering it along the frequency axis. Selected channels of EEG signal are decomposed by Multivariate SynchroSqueezing Transform to capture the frequency components and its amplitudes at any specific time slot. Then, singular value of the matrix is extracted by SVD as the feature vector. Then the most contributing features are selected using AdaBoost ensemble of decision tree classifiers which also leads to a reduction in feature dimensionality. And this reduced feature model acts as input to AdaBoost classifier in One against All (OAA) strategy to discriminate the emotional states in both two dimensional and three-dimensional model comprising Valence, Arousal and Dominance. Experimental results on DEAP dataset show that the proposed method yields classification accuracy of 97% for 2D emotional model and gives better performance than the state-of-art systems by nearly 7%.

S. T. Veena (✉)
Department of CSE, Kamaraj College of Engineering and Technology, Virudhunagar, India
e-mail: veena_st@yahoo.com

M. N. Sumaiya
Department of ECE, Dayananda Sagar Academy of Technology and Management, Bangalore, India
e-mail: drsumaiyamn@gmail.com

D. J. Hemanth (ed.), *Human Behaviour Analysis Using Intelligent Systems*,
Learning and Analytics in Intelligent Systems 6,
https://doi.org/10.1007/978-3-030-35139-7_9

Keywords Human-computer interaction · EEG · Singular value decomposition · Multivariate SynchroSqueezing Transform

1 Introduction

In engineering and medical applications, practical signals are non-stationary, in general, they can change very often. For human-machine interaction, brain activity signal has to be interpreted to make machine understand about the state of human emotions and intensions. One such non-stationary, non-invasive signal is the Electroencephalogram (EEG) which is generated by neuronal activity in the brain. EEG is used to extract the information about the various states of the brain. Recently, a large number of researchers have turned their attention to EEG signal in order to detect human emotions. Hence, human emotion classification using EEG signal has become a popular and challenging research area. However, emotion classification is still a challenging task due to complex neuronal structure of the brain.

Human emotional states are represented in two dimensional (2D) parameter space by valence and arousal scores, which range from negative to positive. Arousal state signifies activity and inactivity in the range of high to low value for excited to bored emotion. While valence score is positive for pleasant, happy emotion, it is negative for unpleasant, sad emotion. These scores are estimated by analysing spectral properties of EEG signals. Thus, researchers reported that power spectral density (PSD) in alpha band (8–13 Hz) indicates the inactivity brain state whereas beta band (14–30 Hz) indicates the active brain state. Generally, arousal level estimation is based on PSD band ratio of the symmetric EEG channels, while valence score estimation is based on asymmetric properties of the EEG signal in the left and right hemisphere. Numerous literatures are available with the use of low/high arousal; low/high valence values [1, 2]. Plutchik [1], proposed classification of eight emotional states such as joy, sadness, fear, anger, surprise, disgust, anticipation and acceptance for emotion classification. Ekman et al. [2] included a greater number of emotions such as shame, guilt, relief, pleasure, satisfaction, pride, contempt, embarrassment, contentment and amusement.

Further, third emotional dimension called dominance is included with valence, arousal dimensions to make more emotional labels in 3D valence-arousal-dominance (VAD) space. This VAD ranges from a sentiment being in control to a sentiment being controlled [3, 4]. But there is a limited work done using VAD space for emotion labelling. In [5–9], VAD dimensions are used to recognize eight emotions by using statistical and high order crossing features. In [10], peripheral psychobiological measurements are evaluated. In [11], authors presented VAD based literature by utilizing regularized extreme machine learning algorithms. In [12, 13], authors used VAD model to detect the emotions from psychobiological signals by using multiresolution approach.

As EEG features are not available directly, emotions are classified through feature extraction from EEG signal. Therefore, effective features play an essential role in emotion recognition yields high accuracy in human emotion classification problem. In addition, to the time domain features, the features from the frequency domain features also have to be dealt with. Say, Empirical Mode Decomposition (EMD) is one of the adaptive time-frequency analysis methods suitable for processing non-stationary and non-linear signals. It decomposes a signal into its intrinsic mode functions without leaving the time domain [13, 14]. This method represents the signal as superposition of a trivial quantity of segments, each of which can be considered as harmonious locally, with moderate changes of their amplitudes and frequencies. The nonlinear characteristic of the EMD method gives a conventional representation possibility. To some extent, however, the utilization of EMD for nonlinear signals has undermined some problems such as the so called "mode mixing" and "aliasing" problems [15]. Recently, EMD-based emotion detection strategies have been presented utilizing standard single and multiple channels of EEG signals [16, 17]. In addition, multivariate empirical mode decomposition (MEMD) has lately been utilized in emotion recognition research [18, 19]. Even though EMD has few advantages for linear and non-stationary [13] data, its mathematical implications are complex to understand. Since EMD origins local changes to the signal, causes loss of signal. In any case, EMD approach tends to be ¨mode mixing¨ which stems from the overlaid intrinsic mode function (IMF) spectra and ¨aliasing¨ which stems from the sub-Nyquist extreme sampling. Even though MEMD has certain favourable applications in dealing with multivariate non-stationary signals, it has a mode-mixing tendency [20, 21].

SynchroSqueezing Transform (SST) is a time-frequency analysis technique [21] which is a blend of wavelet decomposition and reassignment method better than S-Transform [22] and Short-Time Fourier Transform (STFT) [23]. This makes SST a good replacement approach for EMD. The aim of SST is to decompose the signals into some components with time-varying oscillatory features. This is achieved by reassigning the energies of STFT or continuous wavelet transform (CWT), or any similar time-frequency approach, such that the subsequent energies of coefficients are intense around the instantaneous frequency curves of the modulated oscillations. The frequency reassignment step of the procedure builds up the important localization of signal components in time-frequency domain. Moreover, with the development of low-cost multichannel sensors, multivariate extensions of time-frequency algorithms are required to deal with inter-channel dependencies [24]. Thus, SST presents a compact time-frequency representation of multichannel signals [25–27]. On the other hand, the modulated multivariate oscillation model based Multivariate SynchroSqueezing Transform (MSST) used in our study characterizes the compact representation of multivariate signals as a single oscillation structure that captures the attributes of the multivariable signal. In fact, the MSST algorithm divides the time-frequency domain into sections to define a series of modulated oscillations that are common to the data channels in the multivariate data, inspired by the Multivariate EMD algorithm. Considering a multivariate extension of the SST algorithm, a multivariate time-frequency representation is developed in which modulated multivariate oscillations are defined [24].

In [28], emotion recognition method is presented using Multivariate SynchroSqueezing Transform (MSST) of multichannel EEG signals. Using linear support vector machine (SVM), MSST features predict emotions with high accuracy also a comparison with its univariate version is presented along with. It is tested with only on DEAP dataset; more emotions can be facilitated with additional bio signals has been suggested. In [29], time-frequency (TF) representation of EEG signals is proposed for emotional state recognition. Feature extraction and reduction of the high dimensional 2D TF are carried out without losing component.

Generally, the following issues were found in literature. The first issue is the non-stationary nature of EEG signals. Second the dimensionality issue. Third is the right choice of EEG channel. Finally, the choice of the classifier is also an issue. All these issues have been addressed by the proposed system where carefully hand-picked channels are subjected to a compact TF representation by MSST and the features are narrowed down by SVD. Further dimensionality reduction is done by the model of the classifier itself. A fine-tuned classifier is then used to get better results. The rest of the paper is organised as follows: Sect. 2 explains the materials or the database, the basics of the transforms used and the overall methodology. Section 3 details the experimental setup and the results are discussed. Section 4 concludes the paper with suggestions for future enhancements.

2 Materials and Methodology

2.1 Database

The database for emotion recognition analysis using physiological signals (DEAP) [30] is used in this work. It is based on valence-arousal-dominance emotion model and is publicly available. It consists of 32 participants multichannel EEG recordings. For each participant, forty 1 min long music videos were shown and the EEG signals from brain of the participants were recorded while observing these video clips. Expected arousal and valence values compass as large as possible. All the signals were recorded with 512 Hz sampling rate. In pre-processed version, EEG signals were down sampled to 128 Hz.

2.2 SynchroSqueezing Transform (SST)

A non-stationary signal is a signal whose spectral characteristics change with time. To analyse these signals, high quality time-frequency methods are required to describe its time-frequency characteristics. By using these transforms, effective features are selected to preserve the discriminant information from signal into small unique digital value. Present technologies provide efficient ways of such representing signals. An efficient representation followed by simple operations yield good results in many signal processing applications i.e. denoising, compression and prediction. They can give insight into the complex structure of the signal consisting of several components. SynchroSqueezing Transform (SST) is a post-processing technique based on the continuous wavelet transform that generates highly localized time-frequency representations of nonlinear and non-stationary signals. Generally the time-frequency algorithms, such as the short-time Fourier transform (STFT) and continuous wavelet transforms (CWT) have a shortcoming of linear projection which is mitigated in SST. The SST reassigns the energies of these transforms, such that the resulting energies of coefficients are concentrated around the instantaneous frequency curves of the modulated oscillations. As such, SST is preferred over the recently introduced empirical mode decomposition (EMD) algorithm [18]. Though it depends on the EMD by generating localized time-frequency representations, at the same time, it provides a well understood theoretical basis [6]. Signals are represented with time varying amplitudes and frequencies are described by the simple mathematical model as

$$y(t) = A(t)\cos(\emptyset(t)) \tag{1}$$

where A(t) and $\emptyset(t)$ are respectively the instantaneous amplitude and phase, and are termed the canonical pair [7]. The Hilbert transform is applied to the original signal, to form the analytic signal $y_A(t)$ as,

$$y_A(t) = A(t)e^{j\emptyset(t)} = y(t) + iH(y(t)) \tag{2}$$

where *H{.}* is the Hilbert transform operator, and *i* is the imaginary component. The analytic signal $y_A(t)$ is complex valued and admits a unique time-frequency representation for the signal $y(t)$, based on the derivative of the instantaneous phase, $\emptyset(t)$. Recently, the concept of univariate modulated oscillation has been extended to the multivariate case, in order to model the joint oscillatory structure of a multichannel signal, using the well understood concepts of joint instantaneous frequency and bandwidth. For multichannel signal $y(t)$, a multivariate analytic signal is represented by constructing a vector at each time instant as,

$$y_A(t) = \begin{bmatrix} A_1(t)e^{j\emptyset_1(t)} \\ A_2(t)e^{j\emptyset_2(t)} \\ A_3(t)e^{j\emptyset_3(t)} \\ A_4(t)e^{j\emptyset_4(t)} \\ . \\ . \\ . \\ . \\ A_n(t)e^{j\emptyset_n(t)} \end{bmatrix} \tag{3}$$

where $A_n(t)$ and $e^{j\emptyset_n(t)}$ represent the instantaneous amplitude and phase for each channel $n = 1 \ldots N$. The continuous wavelet transform is a projection-based algorithm that identifies oscillatory components of interest through a series of time-frequency filters known as wavelets. A wavelet $\psi(t)$ is a finite oscillatory function, which when convolved with a signal $y(t)$, in the form

$$w(a,b) = \int a^{-1/2} \frac{\psi(t-b)}{a} y(t) \tag{4}$$

gives the wavelet coefficients $w(a,b)$, for each scale-time pair *(a, b)*. In this way, the wavelet coefficients in (4) and can be seen as the outputs of a set of scaled band pass filters. The scale factor '*a*' shifts the band pass filters in the frequency domain, and also changes the bandwidth of the band pass filters. Therefore, the energy of the wavelet transform of a sinusoid at a frequency ω_s will spread out around the scale factor as $a_s = \omega_\psi/\omega_s$, where ω_ψ is the centre frequency of a wavelet, while the energy of the original frequency ω_s is spread across a_s. Thus, the estimated frequency present in those scales is equal to the original frequency ω_s. Consequently, the instantaneous frequency $w_X(a,b)$ can be estimated as

$$w_X(a,b) = -iw(a,b)\frac{\partial w(a,b)}{\partial b}$$

for each scale-time pair *(a, b)*. The resulting wavelet coefficients that contain the same instantaneous frequencies can then be combined using a procedure referred to as SST. For a set of wavelet coefficients $w(a,b)$, the SST is given by

$$T(w_l,b) = \sum_{a_k:|w_X(a_k,b)-w_l| \leq \Delta w/2} w(a_k,b)a^{-3/2}\Delta a_k \tag{5}$$

where ω_l are the frequency bins with a resolution of $\Delta\omega$. The SST is shown to reconstruct univariate modulated oscillations of the form in (1), as follows

$$y(b) = \Re\left[R_\psi^{-1}\sum_l T(w_l,b)\Delta w\right] \tag{6}$$

where $R_\psi = \frac{1}{2}\int_0^\infty \hat{\psi}^*\left(\xi\frac{d\xi}{\xi}\right)$, is the normalization constant and $\hat{\psi}(\xi)$ is the Fourier transform of the mother wavelet *(t)*.

2.2.1 Multivariate SynchroSqueezing Transform (MSST)

The modulated oscillation model, which provides development of modern time-frequency algorithms, gives an expressive representation in the time-frequency domain. To recognize oscillations common to multiple data channels, this concept is expanded for multivariate signals by presenting a multivariate extension of the SST [6]. Given a multivariate data like EEG, the process starts by applying SST to all channels. At this step, SST coefficients are obtained for all channels. After this point, the rules of multichannel dependency of the modulated oscillation model are employed. In order to identify a set of mono-component signals from a given multivariate signal, the time-frequency domain is separated into K Frequency bands $\{w_k\}$, $k = 1, \ldots K$. Then, the instantaneous amplitudes and frequencies which exist within these frequencies are calculated by computing the joint instantaneous frequency of each oscillatory scale across the channels. Thus, multivariate instantaneous frequency and instantaneous amplitude are determined respectively. Finally, the MSST coefficient is determined by employing multivariate instantaneous frequency and instantaneous amplitude.

2.2.2 Multivariate Bandwidth Estimation

MSST decomposes time-frequency domain by using its multivariate bandwidth. In each frequency scale, bandwidth is estimated. The same procedure is applied for all the channels in order to estimate multivariate bandwidth. The joint instantaneous frequency is calculated by taking the power weighted average of the instantaneous frequency of all the channels. Similarly, multivariate bandwidth employs the same concept that the joint instantaneous bandwidth representing the normalized error of the joint instantaneous frequency estimate with respect to the rate of change of the multivariate analytic signal $y_A(t)$. Consequently, the joint instantaneous frequency recognizes the incorporated oscillatory aspect of the multivariate signals given as:

$$w_p(t) = \frac{\Im\{y_A^H(t)\frac{d}{dt}y_A(t)\}}{|y_A|^2} = \frac{\sum_{n=1}^{N} A_n^2(t)\emptyset_n'(t)}{\sum_{n=1}^{N} A_n^2(t)} \tag{7}$$

Here, the symbol "$\emptyset_n'(t)$" signifies the instantaneous frequency for each channel and "$\Im$" represents the imaginary part of the complex signal. The joint instantaneous bandwidth finds the variations of the multivariate oscillations for each channel utilizing the joint instantaneous frequency and is given as:

$$F_p(t) = \frac{\left\| {}^{d}/_{dt} y_A(t) - i w_p(t) y_A(t) \right\|}{y_A(t)} \tag{8}$$

Therefore, the joint instantaneous bandwidth signifies the normalized error of the joint instantaneous frequency evaluation with respect to the rate of change of the multivariate analytic signal $y_A(t)$. Applying (3) in (7) results in the expression for the squared instantaneous bandwidth given by:

$$F_p^2(t) = \frac{\sum_{n=1}^{N} A_n'(t)^2 + \sum_{n=1}^{N} A_n^2(t)\left(\emptyset_n'(t) - w_p(t)\right)^2}{\sum_{n=1}^{N} A_n^2(t)} \tag{9}$$

Therefore, the global moments of the joint analytic spectrum can be represented through the joint instantaneous frequency and bandwidth. The joint mean frequency and the joint global second central moments (multivariate bandwidth squared) are taken as the first and second global moments. Hereby, the joint analytic spectrum is given as:

$$S(\omega) = \frac{1}{E}|y_A(\omega)|^2 \tag{10}$$

where, $y_A(\omega)$ is the Fourier Transform of $y_A(t)$ and the E is the total energy of the frequency domain considered as:

$$E = \frac{1}{2\pi}\int_0^\infty |y_A(\omega)|^2 d\omega \tag{11}$$

The joint analytic spectrum and the total energy make possible to describe the joint global mean frequency specified as the first moment of the joint analytic spectrum as follows:

$$\bar{\omega} = \frac{1}{2\pi}\int_0^\infty \omega S(\omega) d\omega \tag{12}$$

The joint global second central moment (the multivariate bandwidth squared) estimates the spectral deviations of the joint analytic spectrum from the joint global mean frequency, and is equated as:

$$\bar{\sigma}^2 = \frac{1}{2\pi}\int_0^\infty (\omega - \bar{\omega})^2 S(\omega) d\omega \tag{13}$$

Therefore, the global moments of the analytic spectrum can be recognized with the moments of joint instantaneous frequency and bandwidth as:

$$\bar{\omega} = \frac{1}{E}\int_{-\infty}^{\infty} \|y_A(t)\|^2 w_p(t) dt \tag{14}$$

$$\bar{\sigma}^2 = \frac{1}{E}\int_{-\infty}^{\infty} \|y_A(t)\|^2 \sigma_p^2(t)dt \tag{15}$$

Here, $\sigma_p^2(t)$ is the joint instantaneous second central moment (the multivariate bandwidth squared) depicted as follows:

$$\sigma_p^2(t) = F_p^2(t) + \left(w_p(t) - \bar{\omega}\right)^2 \tag{16}$$

The multivariate bandwidth squared, $\bar{\sigma}^2$, relies upon both the joint instantaneous bandwidth, $\sigma_p^2(t)$, and the deviations of the joint instantaneous frequency from the joint global mean frequency, $\bar{\omega}$. As a start, time-frequency domain is sliced into 2^l frequency bands with equal width on the account of the frequency range as $w_{l,m} = \left[\frac{m}{2^{l+1}}, \frac{m+1}{2^{l+1}}\right]$ where $l = 0, \ldots, L$ represents the levels of the frequency bands (default value as L = 5) and $m = 0, \ldots 2^{l-1}$ denotes the index for the frequency band. From this point, the multivariate bandwidth is estimated for each band, then, the bandwidth with power of the signal accounted for is estimated. After decomposing the frequency bin as shown in Fig. 1, the lowest frequency band is divided if power level is higher than the other frequency bands. Therefore, the multivariate bandwidth $B_{l,m}$ within a given frequency band $w_{l,m}$ is divided into the frequency sub-bands $w_{l+1,2m}$ and $w_{l+1,2m+1}$.

SST is applied to each channel to obtain $T(w_l, b)$ coefficients for a given multichannel signal *y(t)* with N channels.

Then the SST coefficient for each channel $T(w_l, b)$ is normalized with the normalization constant, and a given set of oscillatory scales $\{w_k\}$ $k = 1, \ldots K$, for the time-frequency representation, a series of sections along the frequency axis are determined, the instantaneous frequency $\Omega_k^n(b)$ and the instantaneous amplitude $A_k^n(b)$ for each frequency section are determined.

$$\Omega_k^n(b) = \frac{\sum_{w \in w_k} |T_n(w, b)|^2 w}{\sum_{w \in w_k} |T_n(w, b)|^2} \tag{17}$$

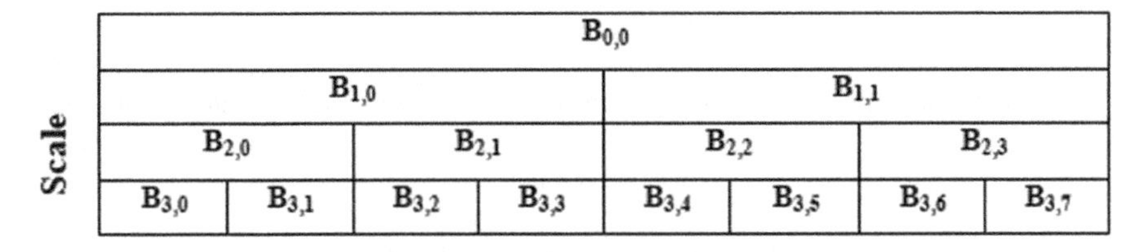

Fig. 1 Decomposed frequency domain with the multivariate bandwidth $B_{l,m}$

$$A_k^n(b) = \sqrt{\sum_{w \in w_k} |T_n(w,b)|^2} \tag{18}$$

Multivariate instantaneous frequency $\Omega_k^{multi}(b)$ and instantaneous amplitude $A_k^{multi}(b)$ are calculated respectively.

$$w_k^{multi}(b) = \frac{\sum_{n=1}^{N}\left(\left(A_k^n(b)\right)^2 w_k^n(b)\right)}{\sum_{n=1}^{N}\left(A_k^n(b)\right)^2} \tag{19}$$

$$A_k^{multi}(b) = \sqrt{\sum_{n=1}^{N}\left(A_k^n(b)\right)^2} \tag{20}$$

The MSST coefficient $T_k^{multi}(w,b)$ is determined as in [8]

$$T_k^{multi}(w,b) = A_k^{multi}\delta\left(w - \Omega_k^{multi}(b)\right) \tag{21}$$

2.3 *Singular Value Decomposition (SVD)*

EEG signal provides a massive amount of information, the percentage of significant signal contribution is less. It is important to extract significant features from a signal. Singular Value Decomposition (SVD) is a reliable orthogonal matrix decomposition method. It becomes more prevalent in signal processing due to its conceptual and stability reasons. Feature reduction/dimensionality reduction is also achieved by SVD which makes the analysis more effective and simple. The MSST coefficients are fed to SVD for extracting significant features from the best MSST representation [31].

$$T = UDV^T \tag{22}$$

where U is the left singular vectors in m $\times$ n orthogonal matrix with the property as below

$$UU^T = U^T U = 1$$

And V is the right singular vectors in n $\times$ n orthogonal matrix such that

$$VV^T = V^T V = 1$$

$D = diag(d_1, d_2, d_3 \ldots d_n)$ with the singular vectors, $d_1 \geq d_2 \geq \ldots \geq d_n \geq 0$. In this paper, the diagonal values are taken into consideration.

2.4 Selection of Channel

Automated Emotion classification using EEG has found controversial issues such as choice of dataset—publically available dataset or privately acquired dataset, domain of features—time, frequency and both time-frequency domains, channel selection—frontal, symmetric. In this paper, the most commonly used database for emotional classification—DEAP database is used. The pre-processed DEAP emotional database consists of 32 × 40 × 32 × 8064 (participants × video × EEG channels × samples) EEG signals which are band passed in the range 4.0–45.0 Hz with sampling rate of 128 Hz. The plot of the EEG signal for the first channel of the first participant viewing first video is given in Fig. 2.

The problem of emotional state identification is more related to neuro science information rather than feature selection. Therefore the channels related to emotional information the symmetric frontal lobes and central lobes are considered for feature extraction as shown in Fig. 3 and Table 1.

2.5 Emotional State Definition

In this paper, two models of emotional states are considered. In the two dimensional model, four emotional states are considered namely happy, sad, fear and calm. The dataset has labels for valence, arousal and dominance rated on the scale of 1–9. The first two parameters attribute to the emotional state predominantly in 2D emotional

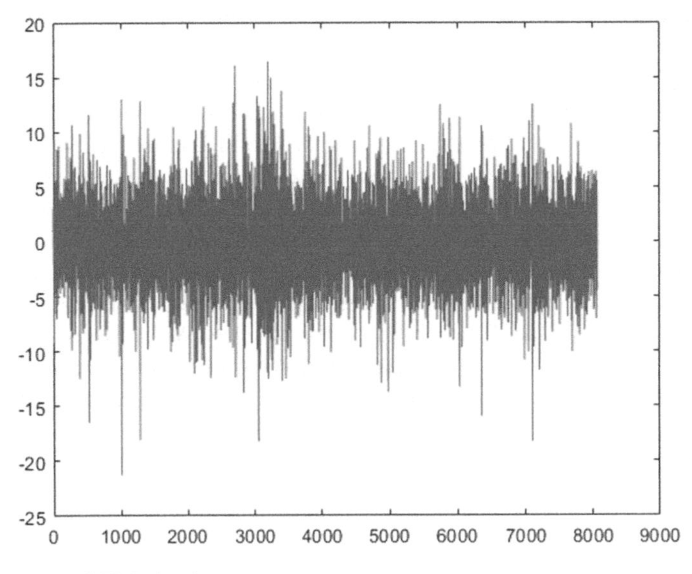

Fig. 2 The plot of EEG signal

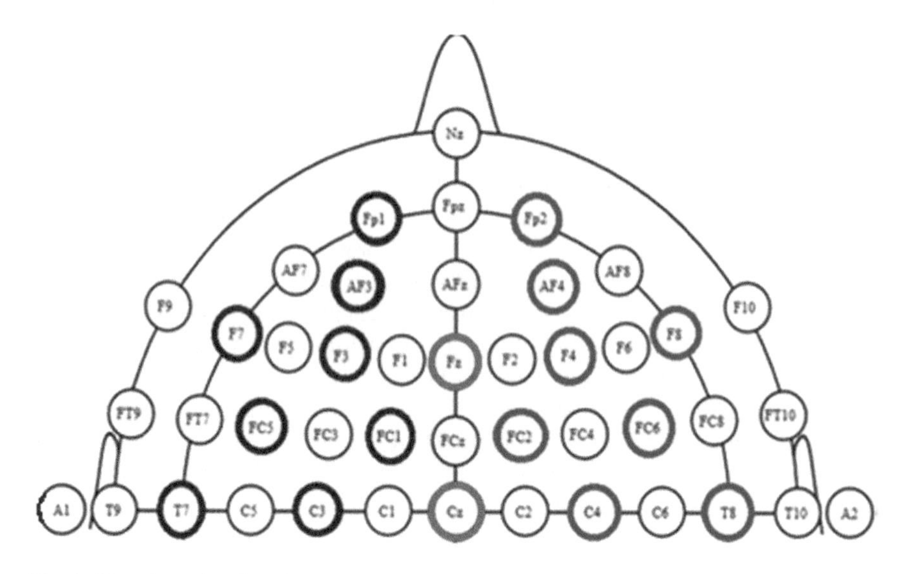

Fig. 3 The channels selected

Table 1 The list of channels selected from the existing channels

Type	Channels
Right frontal	Fp2, AF4, F4, F8, FC2, FC6, T8, C4
Left frontal	Fp1, AF3, F3, F7, FC1, FC5, T7, C3
Central	Fz, Cz

state. The first state NVLA (Negative Valence and Low Arousal—Sadness) has both valence and arousal less than 5. The second emotion NVHA (Negative Valence and High Arousal—Anger) has valence less than 5 and arousal greater than 5. The third emotion PVLA (Positive Valence and Low Arousal—Calmness) has valence greater than 5 and arousal less than 5. The final emotion PVHA (Positive Valence and High Arousal—Happiness) is considered with both valence and arousal greater than five. Thus the four quadrants of the 2D model are considered as the emotional states as in Fig. 4. The three dimensional model consists of states PVHAHD, PVHALD, NVHAHD, NVHALD, PVLAHD, PVLALD, NVLAHD, NVLALD labels for eight emotional states, where P stands for Positive and N for negative, L for Low and H for high, V for valence, A for arousal and D for dominance, This is given in Fig. 5. Human emotion recognition means identification of the emotional states and this calls for the multi classification problem.

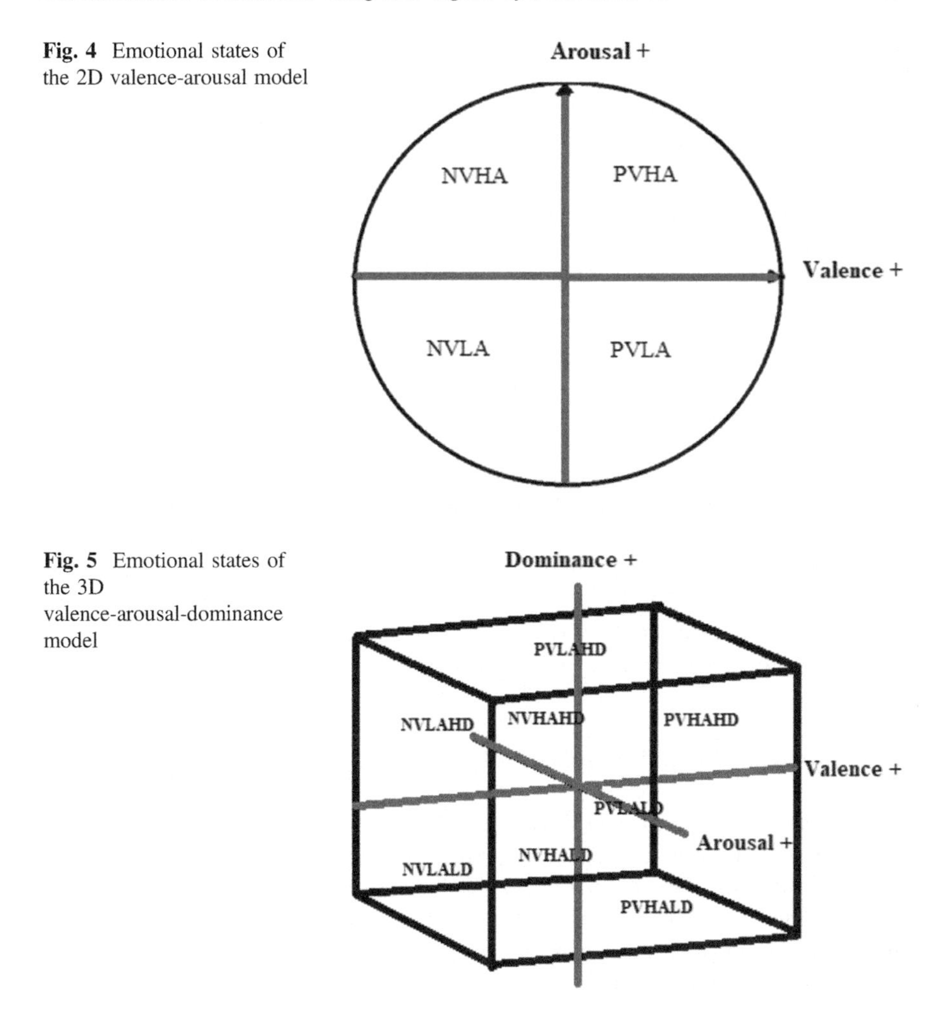

Fig. 4 Emotional states of the 2D valence-arousal model

Fig. 5 Emotional states of the 3D valence-arousal-dominance model

2.6 *Classifier*

The classifiers—Decision Tree, Support Vector Machine (SVM), k-nearest neighbour (k-NN), Multi Neuron Perceptron (MLP), Logistic regression, Linear Discriminant Analysis, Ensemble forest are used in this study. These classifiers are used in their default model from scikit-learn package [34]. A brief description of the classifiers is given below.

2.6.1 Decision Tree

A decision tree is a tree based classifier where the internal nodes are features and a branch represents a decision and the leaf node an outcome. The partitioning of the tree is done recursively based on attribute value. There are various Attribute Selection Measures (ASM) like Information Gain, Gain ratio, Gini index, etc. that are used as the splitting technique. Decision tree classifier in Scikit learn uses Gini index as the default ASM technique. Gini Index is given by

$$\mathrm{Gini}(D) = 1 - \sum_{i=1}^{n} P_i^2$$

where Pi is the probability that an instance (i) in record set D belongs to Class Ci. In case of a discrete-valued attribute, the subset that gives the minimum Gini index is chosen as a splitting attribute. In the case of continuous-valued attributes, the strategy is to select each pair of adjacent values as a possible split-point and point with smaller Gini index is chosen as the splitting point. The other attributes are min_samples_split which is equal to 2 i.e. nodes are expanded until all leaves are pure or until all leaves contain less than 2 samples, and the minimum number of samples required to be at a leaf node is set as 1.

2.6.2 Support Vector Machine (SVM)

A Support Vector Machine (SVM) is a discriminative classifier which classifies classes based on forming a separating hyper plane. For labelled training data, the algorithm finds an optimal hyper plane which can be used to classify new samples. Thus the objective of the support vector machine algorithm is to find a hyper plane in an N-dimensional space (where N is the number of features) that can distinctly classify the new samples. The hyperplane is a line, if the number of input features is 2 and a two dimensional plane, if the number of input features is 3. Many possible hyper planes exist for separating the classes in which case support vectors are used. Certain samples that are closer to the hyperplane, influence the position and orientation of the hyperplane and they are selected as support vectors. SVM works to find support vectors that are adjacent to a hyper plane that has the maximum margin, i.e. the maximum distance between samples of both classes. The varying parameters of the SVM are Kernel, Regularisation and Gamma. Kernel is the representation of support vector as an inner product. Various kernels are linear, polynomial and Radial Bias Function (RBF) kernels. Regularisation parameter (also known as C parameter) helps in lesser number of misclassification of training samples. The optimization will choose a smaller-margin hyper plane for larger values of C such that all the training samples are classified correctly. Conversely, the optimizer with a very small value of C looks for a larger-margin separating hyper plane, even if that hyper plane misclassifies more points. The gamma parameter defines how far the influence of a single training sample is. Low gamma means the samples farther away from the hyperplane are taken into account, while high gamma means the samples close to hyperplane are taken.

2.6.3 k-Nearest Neighbours (k-NN)

It is a non-parametric, instance based learning method used for classification. A sample is classified by a majority class vote of its neighbours (i.e.) for a sample initialise its neighbour is determined by using certain distance measure mostly Euclidean. Then, from its neighbours, the class is assigned that is most common among its k nearest neighbours. Since it is membership of its neighbours it is also known as lazy classifier.

2.6.4 Multilayer Perceptron (MLP)

A perceptron is a linear classifier; that is, it is an algorithm that classifies input by separating two categories with a straight line. Input is represented as a linear equation of feature vector x multiplied by weights w and bias b as $y = w * x + b$. A multilayer perceptron (MLP) is a collection of perceptrons or an artificial neural network which has many layers. Input layer, Output layer and Hidden layers constitute the MLP layers. Input layer receives the signal, while the output layer is the layer that predicts the output from the input. Arbitrary number of hidden layers exists between the two layers. The training is done by repeatedly adjusting the parameters, or the weights and biases, of the model in such a way that error is minimised.

2.6.5 Logistic Regression (LR)

Logistic Regression is a predictive analysis algorithm based on the concept of probability. Logistic Regression uses 'Sigmoid function' also known as the 'logistic function' to predict the class of the samples.

2.6.6 Linear Discriminant Analysis (LDA)

Linear Discriminant Analysis or Normal Discriminant Analysis or Discriminant Function Analysis is used for modelling differences in groups i.e. separating two or more classes. It is used to project the features in higher dimension space into a lower dimension space. Two criteria are used by LDA to create a new axis: Maximizing the distance between means of the two classes and minimizing the variation within each class.

2.6.7 Ensemble Forest

A random forest is an ensemble meta estimator where a fixed number of decision tree classifiers on various sub-samples of the dataset are trained and averaging of the results are used to improve the predictive accuracy and control over-fitting. It employs bagging and bootstrapping concepts.

2.6.8 Proposed Classifier

It is a fact that ensemble classifiers are better performing than a single classifier [32]. In this paper, since it is a multi classification problem and the goal is predict the human emotion state, so a model of ensemble strategy of ensemble classifier is proposed [33]. Therefore a one against all strategy of AdaBoost ensemble bank of decision tree classifiers is employed. AdaBoost algorithm was developed by Freund and Schapire. This algorithm employs a set of weights for the training samples. The objective is to iterate the learning process such that minimum the weighted error on the training set is returned. The change in weights is done iteratively so that more weight is placed on training samples that were misclassified by hypothesis and less weight on samples that were correctly classified. In subsequent iterations therefore AdaBoost constructs progressively more difficult learning problems. And decision tree works on the principle of recursively splitting the features of the source set until further splitting adds no value to the classification result. It is one of the most powerful greedy classification algorithms which are simple to implement. Similarly two strategies of ensemble exists namely the one against one (OAO) and one against all (OAA). In OAA each classifier trains a single class which is the positive class for prediction while all other classes are negative classes, while in OAO k (k − 1)/2 classifiers are trained to detect k classes. Though OAO is more powerful than OAA, OAA is adopted in this paper for its simplicity and relatively better results. Thus this work utilises an ensemble strategy of ensemble classifiers to classify human emotional states of 2D Valence-Arousal model and 3D Valence-Arousal and Dominance model.

2.7 *Feature Extraction and Selection*

The pre-processed feature from 32 participants viewing 40 videos and their EEG from 32 channels are subjected to channel selection which is primarily neuro information related than feature selection. This yields a vector of 1280 $(32 \times 40) \times 18 \times 8064$ from $1280 \times 32 \times 8064$. This information is then converted to a time-frequency representation using MSST transform yielding a vector of $1280 \times 32 \times 352 \times 8064$ which is subjected to reducing dimension using SVD giving 1280×6336 feature vector. Once the TF domain representation is completed, since the dimensionality of features is large, feature selection by employing

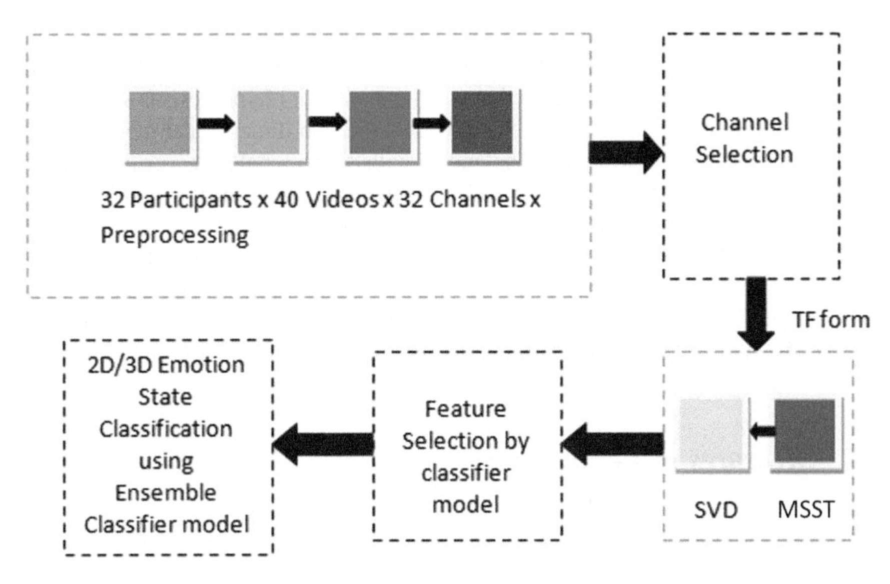

Fig. 6 The proposed method

the AdaBoost classifier hypothesis is done to have the final reduces feature set as 1280 × 48 for 2D model and 1280 × 49 for 3D model. The process of the classification is illustrated in Fig. 6.

The process is also summarized as Algorithm as given below.

Algorithm 1 The Proposed System

MSST-SVD (S)
Input: The pre-processed signal of DEAP database

1. Select the frontal channels including the symmetries and central ones.
2. Apply MSST to each of the selected channel
3. Apply SVD to each of the channel which are MSST transformed
4. Select feature by building a AdaBoost classifier and removing the insignificant features below threshold
5. Apply Ensemble strategy of One against one with an ensemble of AdaBoost Decision Tree classifier
6. Apply the same for 2D and 3D emotional state classification
7. Gather the statistics and Analyse.

3 Results and Discussion

In this paper, the emotional classification of the DEAP dataset using MSST-SVD transform with further feature reduction using the classifier hypothesis is studied. To facilitate the study, two emotional state models the two dimensional model with 4 emotional states and three dimensional model with 8 emotional states are considered. In addition a study of emotion classification using various classifiers is also done. The classifiers—Decision Tree, Support Vector Machine (SVM), k-nearest neighbour, Multi Neuron Perceptron (MLP), Logistic regression, Linear Discriminant Analysis, Ensemble forest and the proposed classifiers are taken into study. These classifiers are used in their default model from scikit-learn package [34]. Cross validation of data in 10 folds is followed with 50–50 train test split with equal class variance.

3.1 Measures of Performance

Three important metrics are used for estimation the correctness of the experiments conducted. They are Classification Accuracy, Confusion matrix and Receiver Operating Characteristics (ROC) curve. The metrics are discussed as below:

3.1.1 Classification Accuracy

Classification Accuracy is the accuracy of the classifier in predict an unknown sample and is measured using

$$\text{Accuracy} = \frac{\text{TP} + \text{TN}}{\text{TP} + \text{TN} + \text{FP} + \text{FN}}$$

where TP is True Positive. It is the number of positive samples which are also predicted as positive. False Negative (FN) are number of positive samples which are predicted as negative, True Negative (TN) are the number of negative samples which are predicted as negative, False Positive (FP) is the count of number of negative samples which are predicted as positive.

3.1.2 Confusion Matrix

A confusion matrix is a matrix which can be used to determine the performance of a classification model on the test data whose truth values are known in advance. It helps in visually see the performance of the classifier. It summarises the count of correct and incorrect classifications along their respective classes. It also gives us insight on the type of errors of classification by the classifier in addition to the accuracy information.

Table 2 Sample confusion matrix

	Class 1 predicted	Class 2 predicted
Class 1 actual	TP	FN Type I error
Class 2 actual	FP Type II error	TN

A sample of the confusion matrix is given in Table 2 and the same can be expanded to multi class classification.

3.1.3 Receiver Operating Characteristics (ROC)

ROC curve is used as a performance metric of a classifier for various thresholds of true positive and false positive rates. ROC is a probability curve and Area Under Curve (AUC) represents the separation measure of the classes and the samples representing them. It is plotted with True Positive Rate (TPR) against the False Positive Rate (FPR) where TPR is on y-axis and FPR is on the x-axis. A curve adjacent to y axis represent a good classification while a straight line or reverse curve tell the negative result or poor classifier. The formula for TPR and FPR are given as.

TPR (True Positive Rate)/Recall/Sensitivity is defined as $\mathrm{TPR} = \frac{\mathrm{TP}}{\mathrm{TP}+\mathrm{FN}}$.

And FPR is defined as $\mathrm{PR} = 1 - \mathrm{Specificity}$, where Specificity is $\frac{\mathrm{TN}}{\mathrm{TN}+\mathrm{FP}}$.

Area under the curve (AUC) is calculated from ROC. A score near 1 implies a good classification model with clear separation space. And near zero represents a poor classifier model.

3.2 Discussion

The result of classification of two dimensional emotional states is given in Table 3 which is derived from confusion matrices obtained from classification. The results show that different classifiers yield different accuracies. Two classifiers were found as most promising namely Decision trees and ensemble forest. The other classifiers were not up to the mark. This may be to the fact that the feature were selected based on a classification model whose hypothesis may be different for other classifiers. Also classifiers need to be fine tuned to the problem rather than arbitrarily using them in their default state.

In the proposed classifier a maximum accuracy of 97% is reached. The PVHA state is much difficult than other states. Since the result of the proposed classifier is significant, confusion matrix for this classifier is given in Table 4. All of 160 samples of PVHA class are predicted as actual label giving a classification accuracy of class PVHA as 100 which is registered in Table 3—last row first column.

Table 3 Classification accuracies of 2D emotional state model with various classifiers using reduced MSST-SVD features

Classifier	PVHA	NVHA	PVLA	NVLA	Average
Decision tree	100	60.99	37.5	93.2	75
SVM	60.0	5.6	22.5	13.75	25.47
k-Nearest Neighbour	36.25	25.63	18.13	13.75	23.44
MLP	1.25	46.25	41.25	10	24.69
Logistic regression	55	40.0	45.0	50.0	47.5
Linear discriminant analyser	43.75	33.75	26.25	13.75	29.38
Ensemble forests	100	0	100	100	80
Proposed classifier	89.37	100	98.75	100	97.03

Table 4 Confusion matrix of 2D emotional state classification with the proposed classifier using reduced MSST-SVD features

No of actual samples of class	No of predicted samples of class				Individual class accuracy
	NVLA	NVHA	PVLA	PVHA	
NVLA	160	0	0	0	100
NVHA	0	160	0	0	100
PVLA	2	0	158	0	98.75
PVHA	17	0	0	143	89.37
Multi class classification accuracy					97.03

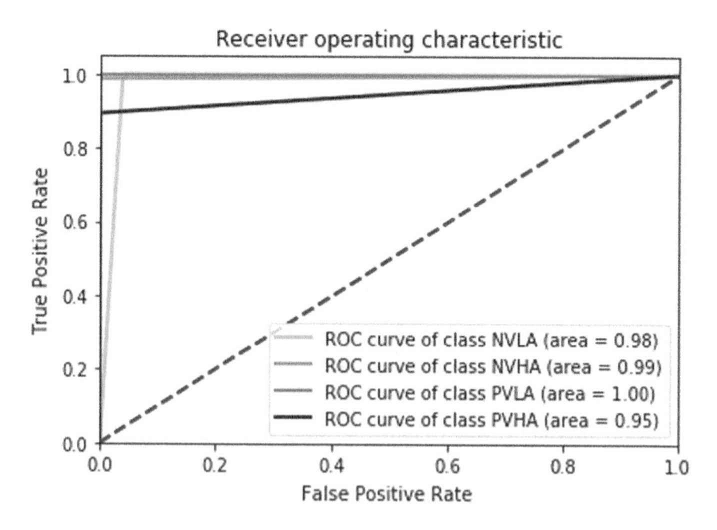

Fig. 7 ROC curve for classification of 2D emotional states using the proposed model

Similarly other class accuracies are calculated and Average accuracy is the overall accuracy of the multiclass classification of a class and is also given in Table 3.

The Receiver operating characteristic (ROC) curve for the same is depicted in Fig. 7.

Table 5 Classification accuracies percentage of 3D emotional state model with various classifiers using reduced MSST-SVD features

Classifier	PVHAHD	PVHALD	PVLAHD	PVLALD	NVHAHD	NVHALD	NVLAHD	NVLALD	Average
Decision tree	99.13	50	97.9	100	61.8	58.82	58.97	100	78.59
SVM	36.25	0	41.25	2.5	6.25	35	0	50	21.41
k-Nearest Neighbour	8.75	0	15	0	15	31.25	6.25	23.75	12.5
MLP	0	0	0	0	0	0	100	0	12.5
Logistic regression	35.0	8.75	18.75	8.75	25.0	18.75	17.5	38.75	21.41
Discriminant analyser	40	1.25	21.25	8.75	15	17.5	3.75	23.75	16.41
Ensemble forests	100	0	80	90.9	77.78	43.75	100	100	84.38
Proposed classifier	87.5	100	93.75	100	100	37.5	100	100	89.84

Table 6 Confusion matrix of 3D emotional state classification with the proposed classifier using reduced MSST-SVD features

No of actual samples of class	No of predicted samples of class								Individual class accuracy %
	NVLALD	NVLAHD	NVHALD	NVHAHD	PVLALD	PVLAHD	PVHALD	PVHAHD	
NVLALD	80	0	0	0	0	0	0	0	100
NVLAHD	0	80	0	0	0	0	0	0	100
NVHALD	20	0	30	0	0	0	0	0	37.5
NVHAHD	0	0	0	80	0	0	0	0	100
PVLALD	0	0	0	0	80	0	0	0	100
PVLAHD	5	0	0	0	0	75	0	0	93.75
PVHALD	0	0	0	0	0	0	80	0	100
PVHAHD	10	0	0	0	0	0	0	70	87.5
Multi class classification accuracy									89.84

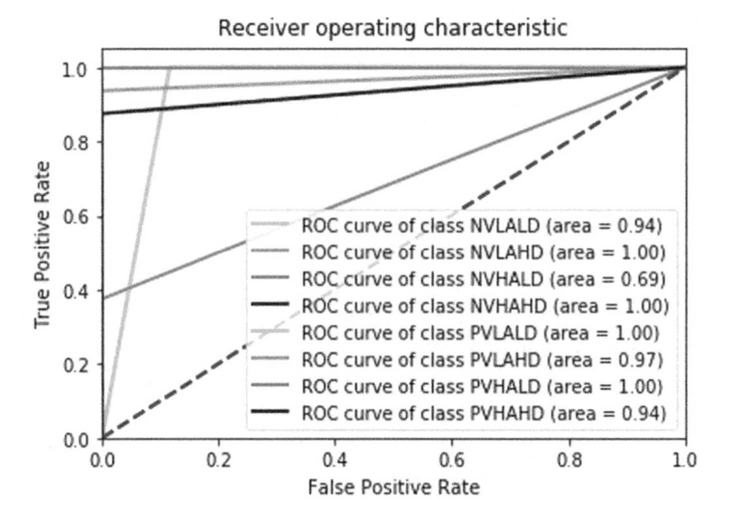

Fig. 8 ROC curve for classification of 3D emotional states using the proposed model

Similarly the classification for the 3D model is given in Table 5 and the confusion matrix is given in Table 6. Here again the performance of the classifiers were relatively in the same order as in 2D classification. The proposed model excelled with 89.84% accuracy. The NVHALD class was identified as the most difficult to identify class. The result less than 2D model is the relative large number of classes the 2D model. The roc curve of one iteration of cross validation is given in Fig. 8.

Comparison of results with existing methods is done and presented in Table 7. As stated earlier, different authors exploit different datasets and classifications for emotional states, therefore a common ground of comparison is hard to find. Most of the work in the area was also devoted to binary classification of the emotional states. The proposed work is compared with multi classification of emotional states since this will lead to human computer interaction and computers understanding and acquiring artificial intelligence. On this context, comparing with the existing systems, the proposed system excels with accuracy greater by at least 7% with minimum features.

Similarly comparing the proposed model for 3D emotional classification with other 3D multi class classification and are tabulated in Table 8. The comparison shows that the proposed method is better than the existing systems.

Table 7 Comparison with existing works for 2D model with multi classification

Work	Dataset, method and classifier	Emotional state	Accuracy rate %
Bajaj et al. (2014) [35]	Real (20), Multi wavelet transform, SVM	4	88.8
Lan et al. (2016) [36]	Real (5), FD, HOC, ICC, SVM	4	49.63
Bajaj et al. (2018) [37]	Real (20), FAWT, KNN	4	86.1
Bajaj et al. (2018) [37]	Real (20), TQWT, ELM	4	87.1
Sachin et al. (In Press) [37]	Real (20) CIF, SVM	4	90.63
Zheng et al. (In Press) [38]	SEED-IV, DNN	4	70.33
Chakladar et al. (In Press) [39]	DEAP (32), CSS	4	82
Chen et al. (In Press) [40]	DEAP (32), EMD, DBN	4	87.32
Ozel et al. (2019) [28]	DEAP (32), MSST-SVD, Ensemble	4	76.33
Proposed method	DEAP (32), MSST-SVD, Ensemble AdaBoost	4	97.03

Table 8 Comparison with existing works for 3D model with multi classification

Work	Dataset, method and classifier	Emotional state	Accuracy rate %
Verma et al. (2017) [41]	DEAP (32), MRA, SVM	9	69.6
Walter et al. (2013) [10]	Real (20), Statistical, ANN	16	70.1
Guendil et al. (2017) [11]	Real (20), Wavelet features, ELM	9	73.43
Ozel et al. (2019) [28]	DEAP (32), MSST-SVD, SVM	8	87.24
Proposed method	DEAP (32), MSST-SVD, Ensemble AdaBoost	8	89.84

4 Conclusion

In this paper, a novel method of human emotion classification using EEG signals in time frequency domain is proposed. The Multivariate SynchroSqueezing Transform (MSST) a recent method can be applied along with Singular Value Decomposition (SVD) to obtain a TF distribution across multiple channels which can be used as compact, well localized feature for emotion recognition. The channels are chosen such that those channels which carry emotional information are selected and the remaining are left out of 32 channels. Further reduction using classifier specific model, produce reduced features of dimension 48 and 49 respectively for 2D (high/low arousal and high/low valence) and 3D emotional state model. The results of the study show that MSST features to extract component information in EEG signals, and classifier specific feature selection to reduce high dimensional 2D TF representation are promising methods and combinations. The proposed method is performed on publicly available DEAP emotional EEG database for comparison. The proposed model classifies with an accuracy of nearly 97% in multi class classification of four states of 2D emotional state model and 89% in 3D emotional eight state model. Thus, TF representation can be successfully used to cover the components in EEG signals for feature extraction, and reduction of the high dimensional 2D TF without losing component information is the most decisive steps of the emotion recognition.

References

1. R. Plutchik, *Emotion: A Psycho Evolutionary Synthesis* (Harper and Row, New York, 1980)
2. P. Ekman, W.V. Friesen, M. O'Sullivan, A. Chan, I. Diacoyanni-Tarlatzis, K. Heider, P.E. Ricci-Bitti, Universals and cultural differences in the judgments of facial expressions of emotion. J. Pers. Soc. **53**(4), 712–717 (1987)
3. A. Mert, A. Akan, Emotion recognition from EEG signals by using multivariate empirical mode decomposition. Pattern Anal. Appl. 81–89 (2018)
4. A.G. Aguinaga, A.L. Ramirez, M.R.B. Flores, Classification model of arousal and valence mental states by EEG signals analysis and Brodmann correlations. Int. J. Adv. Comput. Sci. Appl. **6**(6) (2015)
5. Y. Liu, O. Sourina, EEG-based dominance level recognition for emotion-enabled interaction, in *IEEE International Conference on Multimedia and Expo* (2012)
6. Y. Liu, O. Sourina, EEG databases for emotion recognition, in *International Conference on Cyberworlds* (2013)
7. Y. Liu, O. Sourina, Real time fractal based valence level recognition from EEG. Trans. Comput. Sci. 101–120 (2013)
8. Y. Liu, O. Sourina, EEG-based subject dependent emotion recognition algorithm using fractal dimension, in *IEEE International Conference on Systems, Man, and Cybernetics* (2014)
9. Y. Liu, O. Sourina, M.R. Hafiyyandi, EEG-based emotion-adaptive advertising, in *Humaine Association Conference on Affective Computing and Intelligent Interaction* (2013)
10. S. Walter, J. Kim, D. Hrabal, S.C. Crawcour, H. Kessler, H.C. Traue, Transsituational individual-specific biopsychological classification of emotions. IEEE Trans. Syst. Man Cybern. Syst. **43**(4) (2013)
11. Z. Guendil, Z. Lachiri, C. Maaoui, Computational framework for emotional VAD prediction using regularized Extreme Learning Machine. Int. J. Multimed. Inf. Retr. **6**(3), 251–261 (2017)
12. G.K. Verma, U.S. Tiwary, Multimodal fusion framework: a multiresolution approach for emotion classification and recognition from physiological signals. NeuroImage **102**, 162–172 (2014)
13. G.K. Verma, U.S. Tiwary, Affect representation and recognition in 3D continuous valence-arousal-dominance space. Multimed. Tools Appl. 1–25 (2016)
14. N.E. Huang, Z. Shen, S.R. Long, M.C. Wu, H.H. Shih, Q. Zheng, N.C. Yen, The empirical mode decomposition and the Hilbert spectrum for nonlinear and non-stationary time series anaylsis. Proc. R. Soc. Lond. **454**, 903–995 (1998)
15. D. Ur Rehman, D.P. Mandic, Filter bank property of multivariate empirical mode decomposition. IEEE Trans. Signal Process. **59**, 2421–2426 (2011)
16. A. Khasnobish, S. Bhattacharya, G. Singh, A. Jati, A. Konar, D.N. Tibarewala, R. Janarthanan, The role of empirical mode decomposition on emotion classification using stimulated EEG signals. Adv. Comput. Inf. Technol. **178**, 55–62 (2013)
17. N. Zhuang, Y. Zeng, L. Tong, C. Zhang, H. Zhang, B. Yan, Emotion recognition from EEG Signals using multidimensional information in EMD domain. Biomed. Res. Int. **2017**, 1–9 (2017)

18. C. Guitton, *Emotions Estimation From EEG Recordings* (Imperial College of Science, Technology & Medicine, Department of Electrical & Electronic Engineering, London, 2010)
19. H. Xu, K.N. Plataniotis, Application of multivariate empirical mode decomposition in EEG signals for subject independent affective states classification. Int. J. Commun. **9**, 91–97 (2015)
20. N. Ur Rahman, C. Park, N.E. Huang, D.P. Mandic, EMD via MEMD: multivariate noise aided computation of standard EMD. Adv. Adapt. Data Anal. **5**, 1350007, 25 pp (2013)
21. I. Daubechies, J. Lu, H.T. Wu, Synchrosqueezed wavelet transforms: an empirical mode decomposition like tool. Appl. Comput. Harmon. Anal. 243–261 (2011)
22. Z. Huang, J. Zhang, T. Zhao, Y. Sun, Synchrosqueezing S-Transform and its application in seismic spectral decomposition. IEEE Trans. Geosci. Remote Sens. **54**(2), 817–825 (2016)
23. H.-T. Wu, *Adaptive Analysis of Complex Data Sets* (Princeton University, New Jersey, USA, 2019)
24. A. Ahrabian, D. Looney, L. Stanković, D.P. Mandic, Synchrosqueezing-based time frequency analysis of multivariate data, Signal Process. 331–341 (2015)
25. J.M. Lilly, S.C. Olhede, Wavelet ridge estimation of jointly modulated multivariate oscillations, in *Conference Record of the Forty Third Asilomar Conference on Signals, Systems and Computers* (2009)
26. J.M. Lilly, S.C. Olhede, Analysis of modulated multivariate oscillations. IEEE Trans. Signal Process. **2**, 600–612 (2012)
27. S. Olhede, A.T. Walden, The Hilbert spectrum via wavelet projections. R. Soc. **460**(2044), 955–975 (2004)
28. P. Ozel, A. Akan, B. Yilmaz, Synchrosqueezing transform based feature extraction from EEG signals for emotional state prediction. Biomed. Signal Process. Control **52**, 152–161 (2019)
29. A. Mert, A. Akan, Emotion recognition based on time–frequency distribution of EEG signals using multivariate synchrosqueezing transform. Digit. Signal Process. (In Press)
30. S. Koelstra, C. Muhl, M. Soleymani, J.-S. Lee, A. Yazdani, T. Ebrahimi et al., Deap: a database for emotion analysis; using physiological signals. IEEE Trans. Affect. Comput. **3**, 18–31 (2012)
31. H.C. Andrews, C.L. Patterson, Singular value decompositions and digital image processing. IEEE Trans. Acoust. Speech Signal Process. **ASSP-24**, 26–53 (1976)
32. V. Bolón-Canedo, A. Alonso-Betanzos, Ensembles for feature selection: a review and future trends. Inf. Fusion **52**, 1–12 (2019). https://doi.org/10.1016/j.inffus.2018.11.008
33. X.-C. Yin, K. Huang, H.-W. Hao, DE2: dynamic ensemble of ensembles for learning nonstationary data. Neurocomputing **165**, 14–22 (2015). https://doi.org/10.1016/j.neucom.2014.06.092
34. F. Pedregosa, et al., Scikit-learn: machine learning in python. JMLR **12**, 2825–2830 (2011)
35. V. Bajaj, R.B. Pachori, Detection of human emotions using features based on the multiwavelet transform of EEG signals. Brain Comput. Interfaces **74**, 215–240 (2015)
36. Z. Lan, O. Sourina, L. Wang, Y. Liu, Real-time EEG-based emotion monitoring using stable features. Vis. Comput. **32**, 347–358 (2016). https://doi.org/10.1007/s00371-015-1183-y
37. S. Taran, V. Bajaj, Emotion recognition from single-channel EEG signals using a two-stage correlation and instantaneous frequency-based filtering method. Comput. Methods Programs Biomed. **173**, 157–165 (2019). ISSN 0169-2607. https://doi.org/10.1016/j.cmpb.2019.03.015
38. W.-L. Zheng, A. Cichocki, EmotionMeter: a multimodal framework for recognizing human emotions. IEEE Trans. Cybern. 1–13
39. D.D. Chakladar, S. Chakraborty, EEG based emotion classification using 'Correlation Based Subset Selection'. Biol. Inspired Cogn. Archit. **24**, 98–106 (2018)
40. T. Chen et al., Emotion recognition using empirical mode decomposition and approximation entropy. Comput. Electr. Eng. **72**, 383–392 (2018)
41. G.K. Verma, U.S. Tiwary, Affect representations and recognition in 3D continuous valence-arousal-dominance space. Multimed. Tools Appl. **76**(2), 2159–2183 (2017)